NUCLEAR POWER

The Future of Nuclear Power

JAMES A. MAHAFFEY, PH.D.

Facts On File
An Infobase Learning Company

For Turtle and Lance

THE FUTURE OF NUCLEAR POWER

Copyright © 2012 by James A. Mahaffey, Ph.D.

Facts On File, Inc.
An imprint of Infobase Learning
132 West 31st Street
New York NY 10001

Library of Congress Cataloging-in-Publication Data
Mahaffey, James A.
    The future of nuclear power / author, James A. Mahaffey.
       p. cm. — (Nuclear power)
    Includes bibliographical references and index.
    ISBN 978-0-8160-7654-3
  1. Nuclear energy—Juvenile literature. I. Title.
    QC778.5.M34 2012
    333.792′4—dc23   2011022646

Facts On File books are available at special discounts when purchased in bulk quantities for businesses, associations, institutions, or sales promotions. Please call our Special Sales Department in New York at (212) 967-8800 or (800) 322-8755.

You can find Facts On File on the World Wide Web at http://www.infobaselearning.com

Text design and composition by Annie O'Donnell
Illustrations by Bobbi McCutcheon
Photo research by Suzanne M. Tibor
Cover printed by Yurchak Printing, Landisville, Pa.
Book printed and bound by Yurchak Printing, Landisville, Pa.
Date printed: April 2012
Printed in the United States of America

10 9 8 7 6 5 4 3 2 1

Contents

Preface

Nuclear Power is a multivolume set that explores the inner workings, history, science, global politics, future hopes, triumphs, and disasters of an industry that was, in a sense, born backward. Nuclear technology may be unique among the great technical achievements, in that its greatest moments of discovery and advancement were kept hidden from all except those most closely involved in the complex and sophisticated experimental work related to it. The public first became aware of nuclear energy at the end of World War II, when the United States brought the hostilities in the Pacific to an abrupt end by destroying two Japanese cities with atomic weapons. This was a practical demonstration of a newly developed source of intensely concentrated power. To have wiped out two cities with only two bombs was unique in human experience. The entire world was stunned by the implications, and the specter of nuclear annihilation has not entirely subsided in the 60 years since Hiroshima and Nagasaki.

The introduction of nuclear power was unusual in that it began with specialized explosives rather than small demonstrations of electrical-generating plants, for example. In any similar industry, this new, intriguing source of potential power would have been developed in academic and then industrial laboratories, first as a series of theories, then incremental experiments, graduating to small-scale demonstrations, and, finally, with financial support from some forward-looking industrial firms, an advantageous, alternate form of energy production having an established place in the industrial world. This was not the case for the nuclear industry. The relevant theories required too much effort in an area that was too risky for the usual industrial investment, and the full engagement and commitment of governments was necessary, with military implications for all developments. The future, which could be accurately predicted to involve nuclear power, arrived too soon, before humankind was convinced that renewable energy was needed. After many thousands of years of burning things as fuel, it was a hard habit to shake. Nuclear technology was never developed with public participation, and the atmosphere of secrecy and danger surrounding it eventually led to distrust and distortion. The nuclear power industry exists today, benefiting civilization with a respectable percentage of the total energy supply, despite the unusual lack of understanding and general knowledge among people who tap into it.

This set is designed to address the problems of public perception of nuclear power and to instill interest and arouse curiosity for this branch of technology. *The History of Nuclear Power,* the first volume in the set, explains how a full understanding of matter and energy developed as science emerged and developed. It was only logical that eventually an atomic theory of matter would emerge, and from that a nuclear theory of atoms would be elucidated. Once matter was understood, it was discovered that it could be destroyed and converted directly into energy. From there it was a downhill struggle to capture the energy and direct it to useful purposes.

Nuclear Accidents and Disasters, the second book in the set, concerns the long period of lessons learned in the emergent nuclear industry. It was a new way of doing things, and a great deal of learning by accident analysis was inevitable. These lessons were expensive but well learned, and the body of knowledge gained now results in one of the safest industries on Earth. *Radiation,* the third volume in the set, covers radiation, its long-term and short-term effects, and the ways that humankind is affected by and protected from it. One of the great public concerns about nuclear power is the collateral effect of radiation, and full knowledge of this will be essential for living in a world powered by nuclear means.

Nuclear Fission Reactors, the fourth book in this set, gives a detailed examination of a typical nuclear power plant of the type that now provides 20 percent of the electrical energy in the United States. *Fusion,* the fifth book, covers nuclear fusion, the power source of the universe. Fusion is often overlooked in discussions of nuclear power, but it has great potential as a long-term source of electrical energy. *The Future of Nuclear Power,* the final book in the set, surveys all that is possible in the world of nuclear technology, from spaceflights beyond the solar system to power systems that have the potential to light the Earth after the Sun has burned out.

At the Georgia Institute of Technology, I earned a bachelor of science degree in physics, a master of science, and a doctorate in nuclear engineering. I remained there for more than 30 years, gaining experience in scientific and engineering research in many fields of technology, including nuclear power. Sitting at the control console of a nuclear reactor, I have cold-started the fission process many times, run the reactor at power, and shut it down. Once, I stood atop a reactor core. I also stood on the bottom core plate of a reactor in construction, and on occasion I watched the eerie blue glow at the heart of a reactor running at full power. I did some time in a radiation suit, waved the Geiger counter probe, and spent many days and nights counting neutrons. As a student of nuclear technology, I bring a near-complete view of this, from theories to daily operation of a power plant. Notes and apparatus from my nuclear fusion research have been requested by and given to the National Museum of American History of the Smithsonian Institution. My friends, superiors, and competitors for research funds were people who served on the USS *Nautilus* nuclear submarine, those who assembled the early atomic bombs, and those who were there when nuclear power was born. I knew to listen to their tales.

The Nuclear Power set is written for those who are facing a growing world population with fewer resources and an increasingly fragile environment. A deep understanding of physics, mathematics, or the specialized vocabulary of nuclear technology is not necessary to read the books in this series and grasp what is going on in this important branch of science. It is hoped that you can understand the problems, meet the challenges, and be ready for the future with the information in these books. Each volume in the set includes an index, a chronology of important events, and a glossary of scientific terms. A list of books and Internet resources for further information provides the young reader with additional means to investigate every topic, as the study of nuclear technology expands to touch every aspect of the technical world.

Acknowledgments

I wish to thank Dr. Don S. Harmer, retired professor emeritus from the Georgia Institute of Technology School of Physics, an old friend from the Old School who not only taught me much of what I know in the field of nuclear physics but also did a thorough and constructive technical edit of the manuscript. I am also fortunate to know Dr. Douglas E. Wrege, a longtime friend and scholar with a Ph.D. in physics from the Georgia Institute of Technology, who is also responsible for a large percentage of my formal education. He did a further technical edit of the material. A particularly close, eagle-eyed edit was given the manuscript by my Ph.D. thesis adviser, Dr. Monte V. Davis, whose specific expertise in the topics covered in this work was extremely useful. Dr. Davis's wife, Nancy, gave me the advantage of her expertise, read the manuscript, and saved me from innumerable misplaced commas and hyphenations. The manuscript also received a thorough review by Randy Brich, a most knowledgeable retired USDOE health physicist from South Dakota. He is currently the media point of contact for Powertech Uranium. Thanks to TerraPower, GE-Hitachi, and Atomic Energy of Canada, Ltd., for reading and giving superb feedback on chapters describing their products. Special credits are due to Frank K. Darmstadt, editor at Facts On File; Alexandra Simon, copy editor; Suzie Tibor, photograph researcher; and Bobbi McCutcheon, artist, who helped me at every step in making a beautiful book. The support and the editing skills of my wife, Carolyn, were also essential. She held up the financial life of the household while I wrote, and she tried to make sure that everything was spelled correctly, all sentences were punctuated, and the narrative made sense to a nonscientist.

Introduction

The Future of Nuclear Power has been written as a compendium of advanced designs, practical concepts, and fully developed systems that have yet to be used for the student or the teacher who is interested in seeing where the technology of energy production can go in the next century. This future has already begun, with newly conceived, safer reactor designs being built in the United States and throughout the world. All 104 of our operating nuclear power reactors in this country have become obsolete. The designs, which once seemed exotic and futuristic, are 40 years old, and one by one these vintage Generation II plants will reach the end of productive service in the next 30 years.

The text begins with a description of the currently emerging Generation III power plant designs. Described are the radical new concepts of reactor safety and economy that have been introduced in Japan, Canada, Russia, and the United States. These reactor designs seem fresh and new, addressing some inadequacies in the power plants built in the 1960s, but they may be considered obsolete by the time construction is completed.

Obsolescence moves slowly in nuclear engineering, but these Generation III power plants will be overshadowed by the Generation III+ systems detailed in chapter 2. As the 21st century dawned, so did newer reactor concepts, taking the old but successful boiling water reactors (BWR), pressurized water reactors (PWR), and Canada Deuterium-Uranium (CANDU) systems and turning them into even safer, tougher machines. These plants are capable of withstanding disasters that were not in our imaginations when the age of nuclear power began, and the technology used is beyond the advanced sophistication of the Generation III reactors. The Westinghouse *AP1000*, the Canadian ACR-1000, the GE-Hitachi *economic simplified boiling water reactor (ESBWR)*, the Korean *APR-1400*, and other advanced power plant designs are described in this chapter.

The highly advanced Generation IV reactors in chapter 3 are purposefully being held back so as not to discourage the use of Generation III+ reactors. These reactors go beyond the production of electricity. Anticipating a world that does not depend on petroleum for transportation fuel, the Generation IV reactors are versatile. They can produce hydrogen gas for use in automobiles, and they can use exotic fuels other than uranium.

Challenges are ahead on many fronts, and these practical but underdeveloped reactor concepts are discussed in this chapter as work for the coming generation of technologists.

Chapter 4 details an entirely different development track. The concept of the large, billion-watt power plant may itself be obsolete. Big nuclear reactors can cause big problems, and the 20th century tendency to simply upscale a power plant as needed may not have been the best solution to a growing need for electricity. There is a new push for small, modular reactors. A downsized, fairly inexpensive power plant can be built in a factory and taken to the building site on a truck. Many problems and expenses are eliminated. Several designs that are currently in the works are presented in this chapter. The United States seems to lead in this new avenue, with other countries in close pursuit.

The concept of nuclear-powered transportation is on the leading edge of scientific and engineering thought, and chapter 5 explores its many possibilities. Here, much work will be needed to bring fanciful designs to fruition, and future explorations of the solar system and neighboring star systems will depend exclusively on nuclear power.

There are series of nuclear rocket engines and even jet engines that were designed and tested as far back as the 1950s, primarily without the public's knowledge. These power sources, representing a future of space travel that has already been developed, are studied in chapter 6. Projects named Pluto, Kiwi, Nerva, and Orion were directed toward carrying humans to the outer planets. We may see a need to dust off these concepts in the 21st century.

Chapter 7 introduces the ideas of alternate nuclear economies, in which reactors can run on cheap fuel and produce hydrogen for use in motor vehicles. This may be the ultimate application of the reactors developed under the Generation IV initiatives discussed in chapter 3. As worldwide competition for nuclear fuels arises, the entire energy economy may have to change to meet new challenges.

Finally, chapter 8 covers the future of an extremely important aspect of nuclear technology, the disposal of fission waste. Although the imaginative plans for dealing with spent nuclear fuel may seem to lag behind the reactor designs, work has been underway to address these problems, and many possibilities are revealed in this chapter.

Technical details of the nuclear process are made understandable in this book, through clear explanations of terms and expressions used almost exclusively in nuclear science. Much of nuclear technology still uses the traditional American system of units, with some archaic terms remaining in use. Where appropriate, units are expressed in the international system, or SI, along with the American system. A glossary, a chronology spanning three centuries, and a list of current sources for further reading, research, and Internet access are included in the back matter.

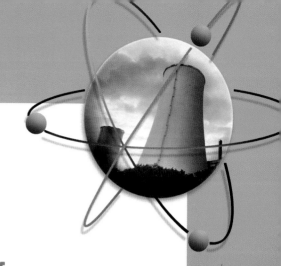

1 Generation III Reactors for the New Century

The Generation I reactors, built in the 1950s, were all unique, hand-built power plants, combining engineering experimentation with real electricity-generating capability. Their purposes were to prove that electricity could be made at a commercial level without using combustible fuel or the strictures of falling water and that the public was not overly endangered by this new source of power. By 1957, the international push for nuclear power was fully underway, at cautious speed. Examples of these pioneering power reactors are the first nuclear plant in the United States, Shippingport in Pennsylvania; the early MAGNOX reactors in Great Britain; the RBMK reactors in the Soviet Union; and the bold attempt at plutonium breeding, the Fermi 1 reactor in Michigan. This first generation of nuclear power proved its points, and much was learned about how to build and how not to build the next generation of power plants. The MAGNOX, the fast breeder, and the RBMK reactor designs were all eventually scrubbed from the list of usable nuclear concepts. The Shippingport design, a pressurized water reactor (PWR), remained standing.

Generation II designs, making use of the lessons learned in Generation I, had matured by the 1970s. Projections for the increase in electrical power demand were high, and the rush to build nuclear power plants was at its peak. Hundreds of plants were either being built or were on order in the United States alone. Industrialized nations all over the world needing more electricity but having few burnable resources, such as France, were planning to replace their entire electricity-generating capacity with nuclear power.

Reactor designs had settled into the following five basic types:

- ❋ The PWR, as designed by Westinghouse, Combustion Engineering, and Babcock and Wilcox
- ❋ The boiling water reactor (BWR), as designed by General Electric
- ❋ The Canadian Deuterium-Uranium reactor (CANDU), as designed by Atomic Energy of Canada Limited
- ❋ The water-water energy reactor *(VVER),* a form of PWR designed in the Soviet Union

By about 1977, the nuclear power rush came to a complete stop, and Generation II designs became frozen in place. The advancement of nuclear technology has remained in this static condition, at least in the United States, ever since. The reason why nuclear power became fossilized in Generation II has nothing to do with safety concerns, public perception, or the involuntary destruction of any power plant. The reason was purely economic. The projected increase in electricity demand did not materialize. There was no need to spend billions of dollars upgrading the electrical system when the existing, coal-fired system was more than adequate. Nuclear plants were expensive to build, costing 10 times as much as a coal plant of similar generating capacity. Uranium was cheap, but coal was cheaper. In 1979, when the Three Mile Island nuclear plant suffered a core meltdown, the nuclear power industry was not affected. There were no new power plant orders to stop. The nuclear construction and support industries in the United States had already shut down. Plants that had been built or were being completed in 1977 were generating power, but there would not be another nuclear power plant construction started in the 20th century.

There was, however, a great deal learned from building and operating more than 100 Generation II power plants. By freezing the level of technology at Generation II for more than 30 years, the nuclear industry was able to observe, compile experiences and operating data, and note aging symptoms of this mature, conservative technology. Even with no demand whatsoever for new nuclear reactors, a new Generation III was devised starting in the 1980s, anticipating a new era for advanced power plants.

Even though these 20th-century ideas and designs are 20 years old, they represent the nuclear power of the future, because not a single Generation III reactor has been built, at least not in the United States. There are some building projects started, using the Generation III designs, and it will take several years before these plants are generating electricity in the United States.

THE GENERAL ELECTRIC ADVANCED BOILING WATER REACTOR

The most successful Generation II reactor in the United States was the Westinghouse PWR. For every BWR that General Electric could sell, Westinghouse sold two PWRs. Babcock and Wilcox and Combustion Engineering were smaller companies, each selling lesser quantities of their own PWR designs. In the dead-quiet years of nuclear stasis after 1977, no reactors were sold in the United States, and the companies big and small found it difficult to maintain the ability to build nuclear power plants.

Babcock and Wilcox of Charlotte, North Carolina, sold its commercial nuclear division to *AREVA,* a multinational industrial conglomerate with headquarters in La Défense, Courbevoie, France.

Combustion Engineering of Stamford, Connecticut, was sold to the *ABB Group,* one of the largest engineering conglomerates in the world, headquartered in Zurich, Switzerland, and jointly owned by interests in Sweden and Switzerland.

In 1999, the Westinghouse Electric Company, famous worldwide for its superb PWR reactors, was acquired by British Nuclear Fuels Limited (BNFL), a company owned by the U.K. Government. Westinghouse had been one of the earliest participants in the development of nuclear power, having built the S2W reactor for the Nautilus submarine in 1954. The Toshiba Corporation and its partners, the Shaw Group and IHI, acquired Westinghouse from BNFL in 2006. Toshiba, headquartered in Tokyo, Japan, is a multinational industrial conglomerate and now owns a great deal of the nuclear technology developed in the United States.

Although there were no prospects for ever selling another BWR in the United States after 1977, General Electric held fast to its successful BWR technology and in 1987 applied for standard design certification from the *Nuclear Regulatory Commission (NRC)* for a new, changed, and improved power plant named the *advanced boiling water reactor (ABWR).*

General Electric managed to retain control of its intellectual property as others were selling theirs, but for the sake of possible overseas business it established a cooperative agreement with *Hitachi*. Hitachi, meaning "sunrise" in Japanese, is headquartered in Tokyo, Japan. Founded in 1910 as an electrical repair shop, it is now the third largest technological company in the world. General Electric formed a joint venture in Japan with Hitachi and Toshiba to sell two of these Generation III ABWR power plants in Japan. The first, at the Kashiwazaki-Kariwa Nuclear Power Plant, started making power in 1996. Four more ABWR units were built in Japan for several utilities, with several more in the planning phase. In 2007, the General Electric Nuclear Energy Division, through a formal alliance with Hitachi, became GE Hitachi Nuclear Energy.

Looking back at the Generation II power plants, there was much to be improved. The plants were complicated and expensive to build and maintain. The control systems were archaic, based on technologies available in the 1950s. The BWR containment structure, designed to prevent any radioactive material from escaping into the environment in a major catastrophe, was particularly troublesome, and its design had gone through several iterations in the Generation II era. In 1981, General Electric was plagued with lawsuits over the BWR containment designs and the expense of upgrading them. The ABWR technology was enhanced in the following six distinctive ways:

※ **Internalized reactor pumps**
A BWR depends on the recirculation of water in the reactor core to control the two-phase flow of the boiling water, and indirectly the fission process. In the Generation II BWR power plants, the two recirculation pumps are located outside the reactor vessel. In the ABWR, the pumps are located inside the vessel through nozzles at the bottom of the reactor pressure vessel. This eliminates any leakage from pipes or pump seals outside the vessel and does away with inlet/outlet penetrations on the vessel for the recirculation pipes.

※ **Combined containment and reactor building**
In the Generation II power plants, the containment structure was a steel pressure vessel, welded together to surround the steel reactor vessel. Its intent was to contain radioactive steam if it escaped, cool it down, and prevent a steam release to the environment under severe circumstances. It was an expensive

Advanced Boiling Water Reactor (ABWR)

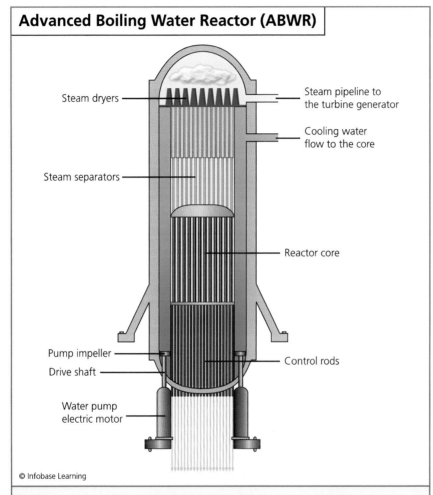

Steam dryers

Steam pipeline to the turbine generator

Cooling water flow to the core

Steam separators

Reactor core

Pump impeller

Drive shaft

Control rods

Water pump electric motor

© Infobase Learning

A schematic diagram showing the internal parts of the advanced boiling water reactor (ABWR) pressure vessel. An important feature is the internally mounted jet pumps used to control the fission process. Unlike the Generation II design, the ABWR pumps are located inside the reactor vessel, with the drive motors outside. The motor shafts stick through the bottom of the vessel.

and complicated way to achieve the containment goal. For the ABWR, the containment structure is part of the reactor building, and it is made of reinforced concrete with a steel liner for leak tightness. It is easier and faster to construct, much less expensive, and more capable for absorbing pressure loads and seismic events, while retaining the good characteristics of the Generation II containment.

※ **Compact reactor building**

Without decreasing the amount of power produced, the Generation III reactor building is smaller than the building used in Generation II power plants. It uses less construction material and is built on a shorter construction schedule.

※ **Modularization**

A problem with the Generation II reactors is that a lot of the complicated pipes and associated equipment had to be assembled on-site as the reactor building was constructed. This was an expensive and inefficient way to build a power plant. In the Generation III design, most of these piping bundles and steam-handling equipment are built in a factory and shipped to the construction site as a module, where they are then connected into the system.

※ **Digital control systems**

In the Generation II reactors, all the controls and data readouts were analog, using needle gauges and pen chart recorders. This antiquated technology was obsolete even as the power plants were being built in the 1970s. The Generation III reactors use computer-based digital readouts, and computer screens give operators a full view of what is going on in the many systems of the power plant. The computer-driven control rooms now include built-in fault testing diagnostics, and the start-up, shutdown, and normal operation of the power plant can now be handled by a system of computers. The control rods on the old system were driven in and out in six-inch (15-cm) steps, but in the new reactors continuous, electric fine motion drives are used, and this is seen as an improvement.

※ **Improved fuel**

The way uranium fuel is manufactured has been improved, reducing the leakage of fission products into the cooling water. This means less radiation exposure for power plant workers servicing the coolant piping and the turbine generator, because BWRs always use a single cooling loop. Some activated products, if released from the fuel into the cooling water, can wind up in the turbine, and when the turbine needs servicing, the workers are exposed to the cooling water. With the new fuel fabrication methods, contamination of the cooling water is greatly reduced.

These changes have resulted in improvements to the successful General Electric Generation II power plants. On May 12, 1997, the NRC issued a final rule certifying the design of the ABWR, and this action made it possible to sell this reactor in the United States. On September 25, 2007, NRG Energy submitted a construction and operations license request to

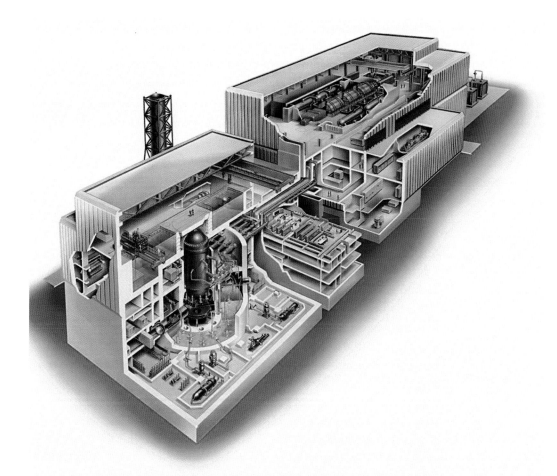

This cutaway picture of the ABWR by GE-Hitachi shows the reactor pressure vessel in the center of the building in the foreground. The reactor sits in the concrete containment structure, which is part of the reactor building. The turbine building is in the upper right, with the turbogenerators on the top floor. *(GE-Hitachi Nuclear Energy, Inc.)*

the NRC for two 1,358-megawatt ABWR nuclear power reactors, to be added to the two existing reactors at the *South Texas Project Electric Generating Station (STPEGS)* near Bay City, Texas. This was the first plan for a Generation III reactor in the United States and the first request for a license in 30 years. The plan was cancelled in April 2011 due to escalating cost estimates.

THE ADVANCED PRESSURIZED WATER REACTOR

Mitsubishi Heavy Industries, Ltd., is headquartered in Tokyo, Japan. The company was started in 1870 as a shipbuilding business in Nagasaki, Japan. It branched out into general heavy industries in 1924, building ships, heavy machinery, airplanes, and railroad cars. By 2007, Mitsubishi was in the nuclear power reactor business, with the sale of two of its Generation III *advanced pressurized water reactors (APWRs)* to the Tsuruga Nuclear Power Plant in the Fukui Prefecture, Japan. The two new reactor units, each generating 1,538 megawatts of electrical power, are expected to be complete by 2016 and 2017.

The APWR has several subtle design advancements that improve on the Generation II PWR safety and economy. Most notably, the reactor core, which is composed of 74,273 rods of uranium fuel, is closely surrounded by a steel neutron reflector. Fewer neutrons are wasted in the fission process, and more are reflected back into the fuel. This results in about 0.1 percent less uranium-235 enrichment required of the fuel and therefore less fuel is required per kilowatt-hour (kWh) of energy produced. The old PWRs had two emergency water tanks, together capable of supplying 100 percent of the coolant water to make up water lost in an accident. The APWR has four water tanks, supplying 200 percent of the makeup water. This improvement led to the elimination of the safety injection system, a complex water makeup scheme requiring power-driven action. The new system requires only gravity to operate.

On December 31, 2007, Mitsubishi submitted the standard design certification application to the NRC. This design is slightly modified to comply with U.S. regulations, and it is called the *US-APWR*. This design uses four improved steam generators, separating the primary and secondary coolant loops. As in all PWR designs, the primary coolant loop contains water under sufficient pressure not to boil into steam. This water heats a second closed loop of water, which boils into steam in the steam generators. The US-APWR makes a drier steam for use in higher efficiency and more delicate turbines. This leads to about a 10 percent increase in energy efficiency. Less fuel is fissioned for the same amount of energy generated.

On September 19, 2008, the Luminant Generation Company filed an application with the NRC for a Combined Construction and Operating License (COL) for two new Mitsubishi US-APWRs at the Comanche Peak Nuclear Power Plant in Somervell County, Texas. Mitsubishi Heavy Industries will own 12 percent of the plant. The high side of the cost estimate is $20.4 billion.

THE ENHANCED CANDU 6 FROM CANADA

There are 29 Canadian-built CANDU reactors in use around the world and another 13 pirated CANDU-type reactors built in India. CANDU reactors are unique in that they use heavy water, or deuterium oxide, as the neutron moderator and primary coolant. The extremely expensive heavy water pushes up the cost of a CANDU reactor, but its efficiency allows the plant to fission a wide variety of inexpensive fuels. Natural uranium, mined out of the ground and purified, can be used. Spent fuel out of an ordinary power reactor can be used. Thorium, or plutonium scavenged out of dismantled nuclear weapons, will fission and make power in a CANDU.

The first CANDU reactor started generating power in Rolphton, Ontario, Canada, in 1962. Since then, there have been incremental improvements of the design. The Generation III CANDU, the *CANDU 6*, has many small but important improvements over the Generation II reactors, as follows:

⚛ **Computers and control systems**
As is the case with all Generation III reactors, the instruments and controls of the CANDU have been upgraded to digital computer process surveillance and automated control. A CANDU reactor can now be started, operated, and shut down by a system of computers.

⚛ **Environmental protection**
The heavy water in a CANDU reactor can activate into tritium, a radioactive isotope of hydrogen, that can make its way into the water supply without special care of the primary coolant. The CANDU 6 continuously extracts tritium from the primary

CANDU 6

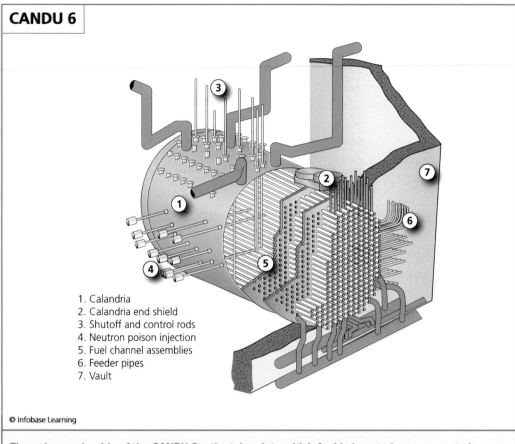

1. Calandria
2. Calandria end shield
3. Shutoff and control rods
4. Neutron poison injection
5. Fuel channel assemblies
6. Feeder pipes
7. Vault

© Infobase Learning

The unique calandria of the CANDU 6—the tubes into which fuel is inserted—are mounted horizontally, so that the fuel may be loaded and unloaded while the reactor is running.

coolant loop, greatly reducing the amount of tritium that can escape into the environment.

❄ Plant maintenance outages

When a plant has to shut down for routine maintenance, this reduces the quantity of power generated over a year's time. A CANDU requires more than the usual amount of maintenance, as the fuel must be cycled almost continuously. As the reactor runs, fuel is inserted into the front of the core and spent fuel falls out the back. The fuel channels wear out and must be replaced on a schedule. To improve the efficiency and the economics of running a CANDU, the Generation III model has an optimized maintenance schedule that is managed by digital computers.

❀ **Plant security**

Nuclear power plants have become a possible target for international terrorists. A nuclear plant is a big target, with much mayhem possible if the nuclear reactor were damaged. Anticipating this problem, the Generation III reactors have stepped up all forms of peripheral security, from fences and armed guards to choke points for cars and trucks inside the facility grounds.

❀ **Fire protection**

A nuclear power plant would not seem a fire hazard. It is made almost entirely of metal and concrete, and these materials will not catch fire. However, a major fire at the Browns Ferry Nuclear Plant in Alabama in 1975 proved that this is not the case. Plastic foam used as a moisture seal around cable penetrations through concrete can catch fire and burn like gasoline, and this can destroy the electrical cable runs. This possibility has been noted

The first CANDU 6 installation at Point Lepreau, Canada; it was the first foreign nuclear power plant to sell electricity to the United States. *(AP Images)*

by the Generation III reactor design teams, and all the new reactors are built using nonflammable materials wherever it is possible to do so. Automated fire extinguishers are in place to put out a fire quickly, if one should develop.

※ **MACSTOR spent fuel storage**
The problem of storing spent fuel for the CANDU has been solved for the Generation III reactors. A new, air-cooled MACSTOR dry storage module for spent CANDU 6 fuel has been developed. It is a short cylinder of concrete, 29.5 feet (9 m) in diameter and 21.5 feet (6.6 m) high. When filled with highly radioactive spent fuel, shielding is sufficient to permit a person to stand next to it without radiation hazard. Over the 50-year lifetime of a CANDU 6 reactor, 275 MACSTOR modules are required to store all the spent fuel.

The first CANDU 6 reactor went into service on February 1, 1983, in Point Lepreau, Canada. Since then, four have been installed in Korea, two in Romania, one more in Canada, and two in China. The last CANDU 6 to go into service was Qinshan 2 in China on July 24, 2003.

THE VVER-1000 AND THE RETURN OF THE RUSSIANS

The disastrous explosion and fire at the Chernobyl Unit 4 power reactor in the Ukraine in 1986 would seem to have stopped the Soviet nuclear power expansion, or surely to have discouraged reactor sales outside the Soviet Union. The Chernobyl *RBMK* reactors were obsolete Generation I designs, devoid of safety features, and the total loss of one reactor spread radioactive dust over most of continental Europe. Unless a country was forced to buy a Soviet reactor, it would seem wise to shop elsewhere. However, this is not what happened. Russia, emerging from the breakup of the Soviet Union in 1990, designed Generation III reactors, selling them to foreign countries and becoming a major supplier of reactor fuel. It was as if Chernobyl had never happened.

The Russian state nuclear company, *Rosatom,* currently supplies 45 percent of the nuclear fuel used to produce commercial electricity in the United States and 17 percent of the nuclear fuel used in the world. By 2025, Russia intends to have 25 percent of the global uranium reactor fuel market. Russia has plenty of highly enriched uranium left over from dismantled nuclear weapons, and it is blended down with depleted uranium-238

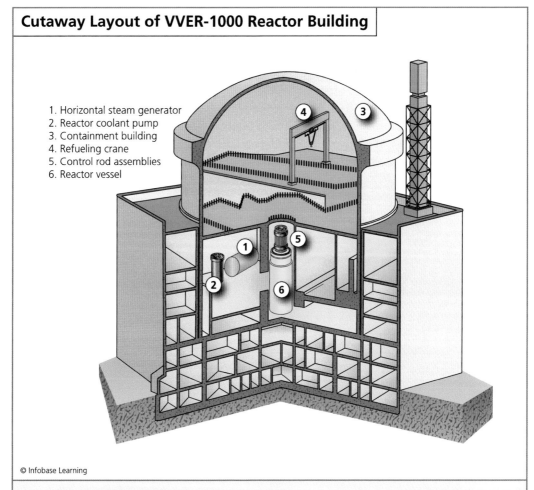

Cutaway Layout of VVER-1000 Reactor Building

1. Horizontal steam generator
2. Reactor coolant pump
3. Containment building
4. Refueling crane
5. Control rod assemblies
6. Reactor vessel

© Infobase Learning

The VVER power plant reactor building can be seen in a cutaway drawing, showing the major parts. This reactor is a departure from the Soviet era RBMK power plants and is similar to Western designs, such as the Westinghouse PWR.

to make 3 percent enriched power reactor fuel. It is an inexpensive source of fuel, and Rosatom competes very favorably on the global uranium market. Russia also has 40 percent of the world's uranium enrichment capability, and appropriately enriched fuel can be made by Rosatom using newly mined uranium.

Rosatom uses inexpensive Russian fuel as an incentive to buy a Russian reactor. Buy the entire power plant, and the customer gets a guaranteed supply of fuel plus fuel reprocessing and long-term storage services. The deal can be irresistible. Rosatom has sold 70 Generation III power plants to Turkey, Bulgaria, Slovakia, Iran, Czech Republic, Ukraine, India, Finland, Armenia, Hungary, and China, with 14 new plants built in Russia. Many more are back-ordered or planned.

This new successful Russian power reactor is the *water-water energetic reactor (VVER)*. Water-water means that the reactor is both moderated and cooled by water, a new concept for Russian reactor design. The current model, the VVER-1000, is a 1,000-megawatt unit, based on the PWR principle first used by Westinghouse for the USS *Nautilus* submarine. In

Russian officials watch the first fuel rod bundle being placed into the new Unit 2 reactor at Volgodansk nuclear power plant in Rostov-on-Don, Russia. The reactor is a VVER-1000/320. *(Matytsin Valery/ITAR-TASS/Landov)*

the United States, commercial power plants were built as upscaled versions of the submarine engine. In Russia, submarine engines are now being built as downscaled versions of the VVER-1000.

The earliest VVERs were the VVER-440 Model V230, built before 1970. An improved design, the V213, was the first Soviet reactor built to any safety standards. The VVER-1000 was developed after 1975, and it incorporates a Western-type containment structure, which is considered to be a very basic safety measure. The current VVER reactors have automatic digital controls, escaped steam suppression, and passive explosion containment systems, which qualify them as Generation III reactors. Unique features include horizontally mounted steam generators, as opposed to the vertically mounted units used on all other PWRs. In addition to generating electricity, the VVER reactors can also supply steam for residential heating and industrial use. The containment building is braced to withstand the destructive force of a jumbo jet crashing into it. The plant should be able to generate electricity for 50 years with some component replacements.

A VVER power plant is almost completely automated. It can start from cold shutdown, generate power, shut itself down, and even refuel itself, but five people must stand there and watch at all times.

The role of the United States, where nuclear power was invented, seems to have been cut back in this new generation of nuclear power development. However, Generation III is just a warm-up.

2 The Improved Generation III+ Reactors

Before the concrete in the foundations for new advanced boiling water reactors (ABWRS) has a chance to harden, these Generation III reactors will be obsolete, outclassed by the GE-Hitachi Economic Simplified Boiling Water Reactor (ESBWR). Just as the industrial world embraced the advanced technology of the Generation III reactors, the Generation III+ had already been designed. All the safety and economic measures incorporated into the reactors that were certified in the 1990s were enhanced further, making it seem advisable to wait for these latest features to be certified before jumping to buy a new power plant. The enticing improvements promised in Generation III+ are worth the delay. By the turn of the 21st century, Generation III+ designs were complete and certification applications were in process at the Nuclear Regulatory Commission (NRC).

A feature in common to all Generation III+ designs is a needed improvement to the emergency core cooling systems, used to maintain a flow of water through the reactor vessel in the event of a major equipment failure and to prevent the fuel from melting. Even when completely shut down, it can take weeks for the fuel in a reactor to reduce temperature to the point where it no longer needs flowing coolant. In the 1960s, when the Generation II power plants were designed, a maximum credible accident was considered to be the sudden breakage of a large steam pipe, and every consideration was given to this theoretical occurrence. Redundant systems of spray heads and extra piping were designed to prevent the fuel from overheating even if the reactor vessel were broken open. It was conventional to move the water to be used for this function using auxiliary pumps, and the pumps were driven by electrical motors.

When a reactor is automatically shut down in an emergency, it no longer produces electricity. In a severe case, there may be no electricity available from other power plants, and in this case there must be an on-site method to generate emergency power. Auxiliary generators, powered by diesel engines, are provided for this event. There are backups to the backup generator. If one fails to start after a shutdown, then another one will start, and if the backup fails to start, there is another one behind it. If no diesel starts automatically, then a bank of batteries will keep the pumps turning until at least one generator can be started. This system of electrically driven emergency coolant pumps plus backup generators was used in all Generation II power plants.

The reliance on diesel generators was a weak point in this generation of reactors. On March 11, 2011, a magnitude 9.0 earthquake shook the main island of Japan. Japan has 17 nuclear power plants, all located on the ocean shores, and most have Generation II reactors. All plants on the eastern shore were shaken by the earthquake without significant damage, but all shut down automatically and quit generating power due to the ground acceleration. Shortly after the earthquake, a tsunami of unprecedented size, 42 feet (13 m), washed over the plants.

All power reactors that had been running started their diesel generators automatically. Auxiliary cooling systems began controlling the temperature in reactor vessels. There were no obvious steam pipe breaks, as had once been feared. At one plant, Fukushima I, the critical diesel generators stopped suddenly as the tsunami inundated the plant. Fukushima I had six boiling water reactors (BWR), three of which had been making power when the earthquake hit, and all but one of 13 emergency generators stalled. Batteries were not able to take the place of the generators for very long. From there, the situation deteriorated over the next few days. Reactors and spent fuel pools overheated. Hot zirconium fuel cladding reacted with steam. Hydrogen gas from the zirconium reaction collected in reactor buildings and eventually exploded, tearing up the last barrier between fission products and the atmosphere. Farmland and the ocean were contaminated with radioactivity, and an entire multiunit power plant was lost, all due to the fact that seawater had contaminated the diesel engines as the tsunami crashed over the plant. Clearly, a dependence on emergency generators to run coolant pumps was a weakness in the Generation II power plant designs. If the generators had kept running at this particular power plant, then there would have been no nuclear disaster to add to the colossal damage caused by this earthquake and tsunami.

The Generation III+ reactors were designed well before the destruction of the Fukushima I power plant, as this weakness had long been recognized. This new series of reactors does not depend on electricity to run the emergency core cooling, so even if the reactor is shut down, all the power lines are down, there are no operating plants on the line, the batteries are dead, and the generators will not start, the reactor can be kept at a safe temperature. These improved designs use passive cooling systems, relying only on gravity and the convective motion of heated water to operate.

THE WESTINGHOUSE AP1000 REACTOR DESIGNS

The first Generation III+ reactor to receive final design approval from the NRC on January 27, 2006, was the Westinghouse AP1000. The review of the design took four years, and the final application document was revision 15. The containment structure of the AP1000, which is the major safety feature against complete system failure, is of a completely new design, and members of the NRC are not entirely convinced that it would survive an earthquake, hurricane, or large airplane collision. Westinghouse is conducting further analysis of the effects of these potential hazards, and the AP1000 is proving to be superior to any Generation II design under rigorous mathematical modeling of these foreseeable accidents.

A major feature of this reactor design is the passive nature of its steam-quenching system. Under the worst possible circumstances, the primary cooling loop of a reactor can be broken. This would result in a steam explosion, as the superheated water in the cooling loop, under pressure, is suddenly exposed to the atmosphere. It would require a severe circumstance, such as an earthquake, to cause such an accident, but it must be considered seriously, and provisions must be made to cool, or quench, the released steam and bring it under control. Primary coolant is likely to contain highly radioactive fission products if the core has exceeded design temperature and has melted the metal cladding on the uranium fuel, and the goal is to keep this material out of the environment surrounding the power plant. All the steam out of the reactor must be contained in the surrounding safety shell, or containment building.

A conceptual picture of the Sanmen Nuclear Power Station in Zhejiang Province. Construction began in 2009, and by 2014 two of the planned six reactors should be running. The reactors are Generation III+ AP1000s. *(Tan Jin/Newscom)*

Generation II power plants have elaborate measures built in to quench the steam, with backup systems in case the primary quenching system fails. These safety systems are complicated, and they require electrical power to run water-spray pumps. Complication detracts from the system's reliability, and an earthquake that breaks the primary cooling pipes could just as well break the water-spray pipes. In the worst case, a power plant could lose all electrical power, and the pumps for the spray-water could not operate. Valves that have remained closed for 30 years have been known to stick.

In the passive core cooling system (PXS) designed by Westinghouse, no electrical pumps are necessary to cool the reactor in an accident, and no human interaction is necessary to initiate the action. The reactor and its primary cooling loop are in a domed enclosure, made of steel. The enclosure is covered by a concrete silo, with an air space between the concrete and the steel. Atop the silo is a water tank, containing enough water to keep the steel enclosure cool for three days. In the event of a steam accident, 20 explosively operated valves blow open, allowing water from the tank to flow through a pipe, directed at the top of the enclosure. Water runs down the sides of the enclosure, and the steam inside condenses into water. Heat transferred to the emergency cooling water causes it to evaporate, and it leaves through a hole in the top of the silo by heat convection. The water-spray is accomplished entirely by gravity, with no pumping action required.

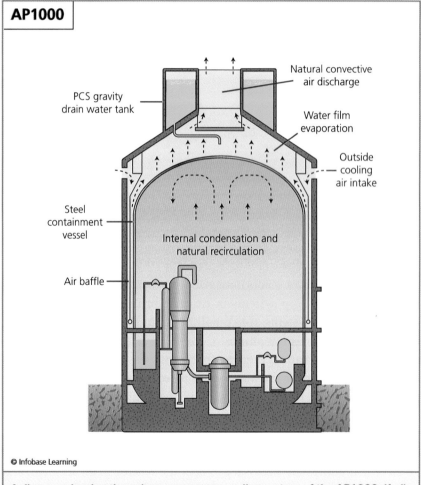

AP1000

Natural convective air discharge

PCS gravity drain water tank

Water film evaporation

Outside cooling air intake

Steel containment vessel

Internal condensation and natural recirculation

Air baffle

© Infobase Learning

A diagram showing the unique emergency cooling system of the AP1000. If all power is lost, the reactor can be kept from melting down by cooling the internal containment structure using water sprayed by gravity from a holding tank in the roof of the reactor building.

Westinghouse engineers expended a great deal of effort simplifying the AP1000, making it more reliable and less expensive to build than Generation II pressurized water reactors (PWRs). They managed to reduce the number of pipes, wires, and valves. In the 1970s, the Generation II Westinghouse PWR was the most popular reactor in the world, and the AP1000 improved on it in the following ways:

- 50 percent fewer safety-related valves
- 35 percent fewer pumps
- 80 percent less safety-related piping
- 85 percent less wiring
- 45 percent less volume in the building

The AP1000 is manufactured in modules, each of which may be transported by rail or river barge, and this shortens the building schedule. For the Generation II plants, too many pieces had to be custom fitted and welded at the building site. The construction period, between the pouring of the first concrete and the first loading of fuel, has been shortened to three years. It took 10 years to build a Generation II power plant.

The Georgia Power Company reached a contract agreement with Westinghouse on April 9, 2008, to install two Revision 16 AP1000 reactors at the Alvin W. Vogtle Electric Generating Plant, near Augusta, Georgia. The plant already has two Generation II Westinghouse PWRs, which began operation on June 1, 1987, with the Unit 2 operating license to expire on February 9, 2049. On February 16, 2010, President Barack Obama advanced $8.33 billion in federal loan guarantees to Georgia Power to construct these two new reactors, and this construction project became the first Generation III+ reactors to be bought in the United States, and the first order for nuclear development since before the Three Mile Island accident in 1979. There are plans for 10 more AP1000s to be built in the United States, but funding for these large projects is uncertain.

THE ADVANCED CANDU REACTOR

A further set of improvements to the CANDU heavy water–moderated power plant design from *Atomic Energy of Canada Limited (AECL)* advances it to the Generation III+ class of reactors. The design of the Advanced CANDU Reactor 1000 (ACR-1000) includes an important change. Since 1962, CANDU power reactors have used only natural uranium, with no enhanced uranium-235 content, and heavy water as an unusually efficient neutron moderator and coolant without parasitic neutron captures. This natural fuel burns up quickly, so to maintain the reactor as a constant power source the CANDU is refueled while it is running. This has always been considered an advantageous arrangement, using inexpensive fuel while avoiding refueling outages. However, using heavy water as the working fluid in the power plant, both as the moderator and the coolant, makes a CANDU reactor an expensive unit. Heavy water costs as much a $270/pound ($600/kg). An American PWR or BWR design uses tap water for the working fluid, at a large cost savings, but it must be run with expensive enriched fuel to compensate for the inefficiency of the light water as a moderator, and the reactor has to be completely shut down and partially dismantled for refueling.

The ACR-1000 has been designed to retain the best traditional features of the CANDU while adopting advantageous features of the PWR and the BWR. The reactor is still heavy water moderated. Horizontal cross-tubes in the core are still used so that fuel may be pushed in one side while spent fuel is collected on the other side. This operation is still done while the reactor is operating at power. However, the moderator and coolant are now separated, never mixing, and the coolant in the ACR-1000 is tap water, just as is used in a PWR or a BWR. This change reduces the heavy water inventory to about 54.7 percent of what is required by a CANDU 6.

The use of light water in coolant channels changes an important fission characteristic. In a light water reactor, such as a PWR, loss of coolant means a loss of neutron moderation, and the reactor is unable to continue the self-sustained reaction in this condition. The same is true in a heavy water–moderated CANDU. Loss of coolant means a loss of moderator, and the reactor automatically shuts down. However, in a reactor using heavy water as the moderator and light water as the coolant, a loss of coolant without a loss of the moderator actually improves the moderating characteristics of the reactor, potentially causing an upward power excursion. This is called a positive coolant void reactivity. For this reason, the ACR-1000 uses low-enriched uranium fuel, with a slightly enhanced uranium-235 content.

The reactivity of the enriched fuel is sufficient to overcome the disadvantages of the moderating characteristics of the light water used in the cooling channels. The majority of all neutrons are still moderated by the heavy water, but there are just enough neutrons slowed down in the coolant and returned to the fuel to maintain criticality. There is not enough

TRIGA: A SAFE RESEARCH REACTOR

Public safety is a primary concern when designing a nuclear power station, and nuclear engineers strive to build nuclear reactors that are absolutely safe. Is this possible? A nuclear power reactor is a complex device, with many opportunities for mechanical breakdowns and accidents. Can the dangerous center of this system, the fission reactor, be designed so that it simply cannot fail? Even if a commercial power reactor is not dangerous to the public, it can be economically dangerous. The core of a reactor, containing the uranium fuel, can overheat under the wrong conditions and melt, ruining a large investment. A research reactor named TRIGA was commissioned on May 3, 1958, and it may be the closest thing ever built to meet this criterion of absolute public and economic safety.

TRIGA is an acronym, made up of Training, Research, Isotopes, and General Atomics, the company that devised and sells these reactors. It was developed as an engineering exercise for a group of young nuclear engineers headed by Edward Teller (1908–2003) in summer 1956, with some key suggestions made by Freeman Dyson (1923–). Both of these men would play interesting roles in a vast range of nuclear projects in the 1950s and '60s. The result of this collaboration is one of the most amazingly successful reactor designs in history. In the words of Frederic de Hoffmann (1924–89), the head of General Atomics, this new reactor design was so completely foolproof that it was "safe even in the hands of a young graduate student."

The key to this melt-proof reactor is the unique formulation of the fuel. The fuel is formed into pellets, which are then assembled into rods and bundles, and these are immersed in ordinary water as a coolant and a moderator, just as in the majority of reactors worldwide. The fuel itself is not metallic uranium, nor is it uranium oxide, with its higher melting point. The TRIGA fuel is a mixture of two solids: uranium and zirconium hydride. The solid fuel pellets are thus in direct contact with a solid moderator. The moderator, which is usually located outside the fuel, is necessary for maintaining fission criticality, for self-sustaining reactions. Without it, the reactor shuts down, and common reactors that overheat tend to boil off the essential water moderator, and this shuts it down, but not before the fuel reaches the melting point. It takes time for the heat of overcritical fuel to travel to the moderator and cause boiling.

Hydrogen is the active ingredient in water that makes it such a good moderator. Hydrogen does combine with uranium, but not in a favorable way. Hydrogen does form a good solid with zirconium, and zirconium is an element with a rare "magic

(continues)

(continued)

nucleus." The nucleus of zirconium is configured so that it cannot absorb neutrons, so it makes an ideal carrier for hydrogen. Mix it with uranium, and there is a moderator in direct contact with the fuel. The heat-transfer time shrinks to zero, and if the fuel overheats, the hydrogen is also overheated. At the temperature of overcritical fuel, hydrogen loses its ability to moderate neutrons. The ability of hydrogen to slow down fast neutrons depends on the fact that the hydrogen is cool, near room temperature. Hot hydrogen cannot slow neutrons down to the desired low energy.

This directly connected uranium-moderator mixture therefore has a desirable property. If the uranium overheats, so does the moderator, and this shuts down the reaction quite promptly. Before any mechanical automatic shutdown system can react, before the operator can leap forward and hit the SCRAM button on the control console, the TRIGA has shut itself down, with no outside influence required. This action is so predictable and so reproducible, the TRIGA is the only reactor that can operate in pulse mode, giving repeated spikes of high neutron activity. This is a valuable feature for several types of research experiments.

More than 60 TRIGA reactors have been licensed for experimental and teaching uses by the NRC. TRIGAs can be found in 20 countries, from Bangladesh to Vietnam, and they are still available from General Atomics, San Diego, California.

excess reactivity in the fuel to run for years without refueling, as in a light water reactor, but there is enough to maintain criticality in the slightly disadvantageous moderator configuration. The ACR-1000 thus has a negative coolant void coefficient and will not sustain fission if coolant is lost accidentally. This new design retains the advantages of continuous refueling capability and is advanced with the low-cost coolant. It will not run with the coolant drained out of it, just as with all previous CANDU designs and light water reactors.

As is the case with all Generation III+ reactors, the ACR-1000 incorporates passive emergency core cooling measures, including an auxiliary holding tank of water that will supply coolant to the reactor by gravity feed even if all electrical power has failed. An advantage of the low-enriched uranium fuel over natural fuel is that a lesser volume of fuel is required to generate the same number of kilowatt hours of energy, and there is less waste material to be handled. The reactor can burn a variety of fuel, including uranium mixed with plutonium or thorium. The fuel is only enriched to 1 or 2 percent uranium-235, whereas a light water–moderated reactor uses 3 to 4 percent uranium enrichment, at higher cost.

The reduced operating and building costs of the ACR-1000 will make it competitive in the international energy market, and its designed lifetime is 60 years. A single unit will generate 1,200 megawatts of electricity. The first one is planned to be in operation in Canada by 2016.

These plans were put on unlimited hold by the AECL in June 2011. The Ontario government, which had agreed to buy two reactors at a cost of $30 billion, cancelled its order because of concern over buying untested nuclear technologies without a guarantee of federal government insurance against cost overruns. The development project had already cost the Canadian tax payers more than $300 million, and much negotiation will be necessary to restart the project so as not to waste this investment.

ECONOMIC SIMPLIFIED BOILING WATER REACTOR

In 50 years of operational experience, the most troublesome components in nuclear power plants have seemed to be water circulation pumps, followed closely by valves. These are electrically actuated units with internal moving parts. With the motor or the actuator being outside the unit and the moving part, such as the pump impeller, being inside the unit, there must be at least one seal that separates the hot, moving water from the outside world. There is always a shaft or a rod that must turn or slide in this seal. At this crucial interface, a leak can develop. The electrical motor or actuator is susceptible to overheating, water damage, fatigue, or simple wear, and the electrical wires can be broken, burned, or short-circuited due to a breakdown of the insulating coating. These components are expensive and difficult to replace, and spares must be kept in an off-site warehouse in case of a failure. Building, running, and paying for a nuclear power plant would be much easier if there were fewer pumps and valves. The optimum power plant of engineers' dreams would have no pumps at all.

On February 14, 2002, the U.S. secretary of energy Spencer Abraham (1952–) unveiled the Nuclear Power 2010 Program as a challenge for companies in the United States to develop new nuclear reactors, anticipating a need for a new pollution-free power source. The bottom-line goal of this program, and the Energy Policy Act that followed in 2005, was to get the price per kilowatt-hour of nuclear-produced energy down to the point where it can compete favorably with other sources of emission-free

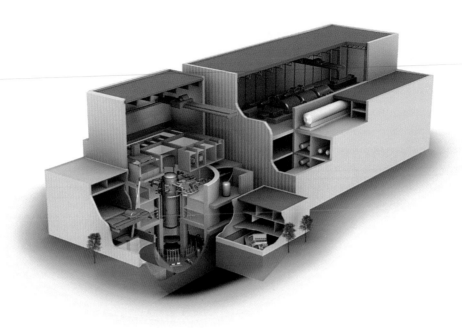

A cutaway picture of the GE-Hitachi economic simplified boiling water reactor *(GE-Hitachi Nuclear Energy)*

power, including windmills. A government/industry cost-sharing plan for the innovation effort was added as an incentive. On August 24, 2005, GE-Hitachi submitted the application for standard design certification to the Nuclear Regulatory Commission for its contribution to the nuclear power program, the economic simplified boiling water reactor (ESBWR).

This Generation III+ design was definitely a long step forward, and the engineers had taken the challenge for reduced costs very seriously. The height of the active core in the reactor vessel was shortened, from 12 feet (3.7 m) in the ABWR to 9.8 feet (3 m). This reduces the pressure drop as cooling water runs vertically through the core, and it allows natural circulation to take the place of pumped circulation. Thermo-siphon action has taken the place of electrical pumps. As the water is heated by the fuel, the hot water rises in the vessel, and this pulls in fresh, cold water through the inlet in the bottom of the vessel. The heated water becomes steam and exits the top of the vessel to turn the turbogenerator. Spent steam, made heavy by condensation into water droplets, falls by gravity into the condenser, where it is further cooled into running water, and it is sucked into the bottom of the reactor vessel. There are no pumps and few valves.

This is not a downsized power plant. It has 1,132 bundles of fuel in a normal-sized BWR reactor core, and it generates 1,600 megawatts of power. The overall plant efficiency for turning heat into electricity is 35 percent. That is quite good for a machine that does not circulate its working fluid by force.

Furthermore, there are no pumps in the passive containment cooling system, used for extreme emergencies in which the reactor vessel has broken open. A very large pool of water, on the same level as the reactor core, surrounds the reactor in a donut shape, or a torus. Excess heat from a reactor breakdown is transferred to the water in this pool. The heat causes the water to evaporate into the air in the large volume of the sealed reactor containment building. The water vapor condenses on the walls of the building and runs back into the pool, with the heat radiated off the sides and top of the building. Using this safety system, the ESBWR can remain safely stabilized for 72 hours without any human action. The probability of a radioactivity release from this reactor into the surrounding environment is much lower than any Generation II power plant.

The building cost of the ESBWR is estimated to be 60 to 70 percent of the cost of building any other Generation III or Generation III+ power plant, and the operating cost may be lower. Construction and operating license applications for GE-Hitachi ESBWRs have been applied for by the Grand Gulf Nuclear Power Station in Port Gibson, Mississippi, the River Bend Nuclear Generating Station in St. Francisville, Louisiana, and the Enrico Fermi Nuclear Generating Station in Frenchtown Charter Township, Michigan. The first application, for Grand Gulf, was made on February 27, 2008. There is some distance to be traveled between a license application and starting up a completed power plant, but in these cases the first step has been taken.

THE EUROPEAN PRESSURIZED REACTOR

In the 21st century, the concerns of nuclear power safety are different from what they were in the 20th century. In the 1970s, as the nuclear power building boom was peaking, the major fear of reactor failure was hypothetical. If a major pipe, such as the steam outlet feeding a steam generator, were to suddenly snap off flush with the reactor vessel, then the cooling water would boil away quickly, leaving the core uncovered and free of cooling. The core would melt as the steam escaped, and the steam would be contaminated with fission products from the wrecked fuel. The steam pressure would break open the building surrounding the reactor, and a plume of radioactive steam would be carried across the countryside on the wind. A great deal of effort was spent ensuring that this would never happen, and the safety measures addressing a major pipe break ran up the cost of building nuclear power plants.

This accident has never happened. The reactor vessel and its pipe intersections are so strongly built, it is unlikely to happen, even under the most severe circumstances. Reactor accidents that render a power plant forever unusable have been much more subtle. A motor-operated valve stuck open, for example, ruined the brand-new Three Mile Island Unit 2 in Harrisburg, Pennsylvania, in 1979. At Fukushima I in Japan, the entire plant was destroyed because the diesel backup generators were flooded with seawater. Accident prediction is a difficult and inexact art.

Now, our worries are about terrorist actions directed against large nuclear power installations. Terrorists have proven that they can take down major structures, such as the World Trade Center in New York City in 2001, without dealing with guarded perimeters. The worst-case nuclear accident is now not due to weak pipes or engineering errors. What is now feared is a large commercial airliner driven directly into a reactor containment building.

The new Generation III+ power plant designed by AREVA is the *European pressurized reactor,* which has the international name of *evolutionary power reactor (EPR).* As is the case with all Generation III+ designs, a major feature of this reactor is that its containment building is double-walled, steel inside concrete, built to survive an airliner crash aimed for the center of the building. Steam-break safeguards are also in place, and they are now quadruple redundant, with four independently exercised steam-quenching measures located in four separate buildings. The geographic separation is designed with the assumption that no terrorist plot could manage to secure as many as four aircraft for an attack. The spent fuel building and the main control room are similarly protected against outside forces. All new Generation III+ plants are also designed to withstand earthquakes. This is a feature that had to be retrofitted at great expense into Generation II plants but is now commonly designed into all nuclear facilities.

The standard design certification application for the EPR was submitted to the NRC on December 11, 2007. The EPR has digital computer controls and process monitoring, as do all post–Generation II designs. On November 4, 2009, the nuclear regulatory commissions in France, Finland, and the United States issued a joint letter to AREVA, citing possible flaws in the EPR digital controls. The safety systems, designed to handle accidents or system breakdowns, were found to be reliant on connections to the operating controls and computers. A safety system must be completely independent of these controls, so that there is a backup system if the main control system falters. For this design to be certified, these systems will have to be redesigned.

The EPR generates 1,600 megawatts of electricity. Its energy-conversion efficiency, from thermal to electrical, is 36 percent, making it 1 percent

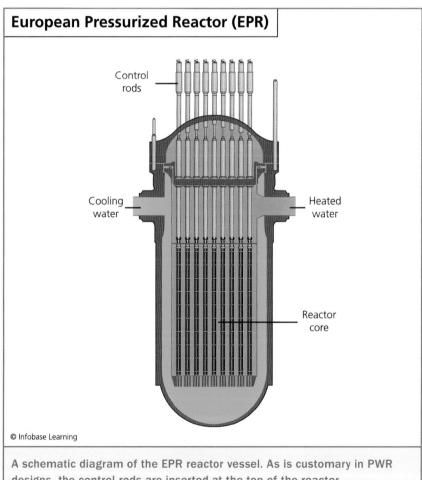

European Pressurized Reactor (EPR)

Control rods

Cooling water

Heated water

Reactor core

© Infobase Learning

A schematic diagram of the EPR reactor vessel. As is customary in PWR designs, the control rods are inserted at the top of the reactor.

more efficient than the simpler ESBWR from GE-Hitachi. An additional safety feature is one that is unique to European and Russian Generation III+ reactors. It is a "core catcher," designed to collect the melted remains of the reactor core as it burns through the bottom of the reactor vessel. This corium area in the bottom of the containment building is cooled by water in a dedicated tank. This feature seems to indicate a lack of confidence in the four independent, redundant emergency core cooling systems.

On February 17, 2005, the government of Finland issued a building permit to Teollisuuden Voima Oyj (TVO) for an AREVA EPR, to be erected at the existing Olkiluoto Nuclear Power Plant on Olkiluoto Island, Finland. It was the first EPR power plant purchased, and it may be the first Generation III+ reactor to be built in the world. Construction began in August, to be completed in 2009. By May 2009, the project was, unfortunately, more than 50 percent over budget and three and a half years behind schedule. Problems in building this new type of reactor surfaced repeatedly, always slowing the progress and making it necessary to rip out

The turbine deck in the Olkiluoto 3 power unit in Eurajoki, Finland. The reactor is an EPR, expected to be operating in 2013. *(Jacques Demarthon/AFP/Getty Images)*

concrete or reject steel forgings and welding jobs. The completion date has been moved forward to 2013.

These difficulties have not made it easy to sell more EPRs, but there is at least one more EPR under construction. On December 6, 2007, concrete was poured for the foundation of the Flamanville Nuclear Power Plant Unit 3 in Manche, France. Completion was planned for 2012. By April 2008, building inspectors began finding cracks in the concrete base of the very important containment structure, and a quarter of the welds in the steel containment liner were found to be unsatisfactory. These findings have led to delays and cost overruns, and the completion date was reestimated to be in 2013.

Further reactor contracts for building the EPR have failed to materialize in Abu Dhabi, United Arab Emirates, China, and the United Kingdom. In April 2009, plans for an EPR at the Callaway Nuclear Generating Station, Missouri, were suspended by the Missouri legislature, and in July 2010, an EPR construction at Calvert Cliffs Nuclear Power Plant, Maryland, was cancelled because of loan guarantee uncertainties. Plans for EPRs are pending in India and Guangdong Province, China.

THE KOREAN APR1400

The rise of South Korea from a small war-torn country in the middle of the 20th century to the economically powerful state it is today is impressive. South Korea has few natural resources but abundant ambition, and it has managed to grow from an agrarian economy to a major technical center of manufacturing and innovation. Korea now has 20 power reactors in four nuclear generating stations, producing 45 percent of its electricity. More are planned, under construction, and being tested.

The push for nuclear power began in 1978, with a small, 556-megawatt reactor built by Westinghouse at the Kori Nuclear Power Plant in Busan, South Korea. The Generation II power plants in South Korea, including four CANDU reactors, were all built under foreign contracts, but the value of a domestic product was seen early. Since 1995, South Korean power plants have been built using indigenous technology. Local industrial expertise formulated a *Korean standardized nuclear plant (KSNP)* design. Between 1998 and 2002, six KSNP plants were completed and are now generating power, using 95 percent native South Korean technology. By 2012, Korea plans to be fully self-sufficient, able to handle all equipment and fuel design and manufacturing.

Workers clap as the locally built APR1400 reactor vessel is installed in the power plant New Gori 3 in Ulsan, South Korea. *(Yonhap News/YNA/Newscom)*

A standard design is an excellent plan. It is easier and much less expensive to build power plants when they are all alike, with standardized parts, operating procedures, worker training, and fuel handling. A standardized nuclear reactor design has been attempted in the United States. The problem with a standardized design is that, in the world of constantly improving nuclear technology, any standard quickly becomes obsolete.

The evolution of the KSNP began in the United States, with a PWR designed by Combustion Engineering (C-E). The Palo Verde Nuclear Generating Station was built in Wintersburg, Arizona, in 1976. It began producing power in 1988, using three C-E System 80 PWRs, generating a total of 3,200 megawatts of electricity. Although the System 80 was technically a Generation II reactor, it was the last of the old designs, and it was considered to be a better, more economical plan for a power plant. C-E was absorbed by the ABB Group, and no more System 80 reactors were built, but the Duke Power Company in Charlotte, North Carolina, saw merit in the design and developed some improvements.

The enhanced reactor is the System 80+. It has a safety depressurization system, to carefully lower the pressure in the reactor in cases of overheating. A gas turbine is connected at the end of the turbogenerator shaft, to take over the job of producing electrical power if the reactor has failed or is simply down for refueling. It has the usual improvements in emergency core cooling and general heat-removing characteristic of Generation III reactors. The System 80+ design was certified by the NRC on May 20, 1997, and the advanced features of this design were adopted by the Korea Power Engineering Company, Inc. (KOPEC).

Seeing that the design could be improved further, KOPEC reengineered the reactor, giving it a more robust foundation and containment structure, able to withstand an earthquake that would shake a Generation II power plant to pieces. Its first designation was the Korean Next Generation Reactor (KNGR). It was awarded design certification in May 2003 by the Korean Institute of Nuclear Safety and given the name APR1000, a Generation III+ reactor. The first two are under construction at the Kori Nuclear Power Plant and are expected to begin producing power in 2013.

On December 9, 2009, the United Arab Emirates shocked all other applicants, including France and the United States, by awarding a $20 billion contract to Korean Electric Power to build four APR1400 reactors. South Korea is now an exporter of its Generation III+ reactor technology. By 2030, this small country plans to export 80 reactors.

A new VVER-1200 reactor being built at the Leningrad Nuclear Power Plant in the Leningrad Region on July 15, 2010. New reactors will gradually replace the four RBMK-1000 reactors currently being used. *(Babushkin Andrei/Newscom)*

THE RUSSIAN VVER-1200/491 SYSTEM

The Russians have stepped up to the challenge of a Generation III+ reactor design for what is seen as a developing global market, with evolutionary enhancements of the VVER-1000. This new design, the VVER-1200, can now withstand an airliner crashing into the containment building and a major earthquake across the foundation, and these improvements alone qualify it as a Generation III+ reactor.

As is the case with all other Generation III+ designs, the VVER-1200 has passive water-cooling systems in case of complete system failure and a double-walled containment structure, with concrete over steel. Thermal efficiency of the VVER-1000 was only 31.6 percent, but the new design is 36.6 percent efficient. The previous reactor had a 30-year expected lifetime, but to make the VVER-1200 competitive in the world market, it has an improved lifetime of 50 years. The electrical power output has been increased to 1,290 megawatts.

Four units, designated VVER-1200/491, are being built in Russia. Construction began on the first unit at the new Leningrad Nuclear Power

Plant II on October 25, 2008. Two more are being erected at the Novov-oronezh Nuclear Power Plant in Voronezh Oblast, and these reactors are expected to be completed between 2012 and 2017. Hopes are to sell a repackaged reactor, the Modernized International Reactor (MIR-1200), in greater Europe.

These advanced Generation III+ reactors meet the requirements of a newly awakened world, having a greater electrical power demand, less money to spend, and less tolerance for unsafe mechanisms. Unfortunately, these computer-controlled, airplane-resistant reactors will be obsolete by the time they start generating power about 2013. The Generation IV reactors are already being designed.

3 Theoretical Generation IV Reactors

In January 2000, the U.S. Department of Energy (DOE) Office of Nuclear Energy, Science and Technology convened a group of senior governmental representatives from nine countries. The purpose of this meeting was bold and unprecedented. It was to begin discussions on international collaboration for the development of advanced nuclear energy systems. Countries in attendance at this first meeting were Argentina, Brazil, Canada, France, Japan, the Republic of Korea, the Republic of South Africa, the United Kingdom, and the United States.

The conference was named the *Generation IV International Forum (GIF),* and its purpose was to identify concrete targets and begin research and development efforts on a new generation of nuclear power reactors. There was some danger of making the Generation III+ reactors obsolete before their sales contracts were finished due to this Generation IV lineup of advanced reactors. This new development could possibly stall the reemergence of practical nuclear power systems with promises of even better reactors. To prevent this impression, the implementation date of Generation IV was purposefully pushed forward to 2030. Goals set for these nuclear power systems are impressive:

❋ **Sustainability.** These reactors will provide sustainable energy generation that meets clean air objectives and provides long-term availability of the systems and efficient fuel utilization.

The nuclear waste and its long-term stewardship burden will be minimized, thereby improving the protection of the environment and the public health.

❈ **Economics.** The Generation IV power systems will be designed with a clear life cycle cost advantage over all other energy sources. The financial risk of building a Generation IV power plant will be comparable to other energy projects.

❈ **Safety and Reliability.** These reactors will be designed specifically to have the lowest possible probability of reactor core damage, and they will excel in recorded safety and reliability. The need for off-site emergency response will be eliminated.

The intended role of these power plants is more than electrical power generation. Other potential uses are synthetic gasoline and diesel fuel production, hydrogen gas production, seawater desalination, and steam heat production. Enhanced international collaboration will avoid unnecessary duplication of research and development efforts, and this arrangement should reduce the costs of designing these systems.

The GIF Charter was signed by representatives of all participating nations between July 3 and July 25, 2001. The United States signed on July 18, 2001. Switzerland joined the forum and signed on February 18, 2002. The *European Atomic Energy Community (Euratom),* a coalition of western European nations headquartered in Brussels, Belgium, joined the GIF in 2003, and in November 2006 the Russian Federation and the People's Republic of China signed on. Within the GIF Charter is the Framework Agreement, in which participants formally agree to actually develop one or more Generation IV systems. Eight of the 12 participants have signed this agreement: the United States, Canada, Euratom, France, Japan, China, Korea, and Switzerland. Annual meetings of the GIF have occurred since the beginning in 2000, and annual reports have been issued beginning in 2007. These reports provide updates on the GIF's organization and membership, as well as a summary of research and development action for each Generation IV system.

The existence of this organized approach to nuclear power development is encouraging. Its international participation and cooperation show that this discipline may have progressed far beyond the point of the birth of nuclear power, when nuclear explosives and extreme secrecy were prominent concepts. Six advanced reactor concepts are described in the Generation IV technology road map completed in 2002: the very high temperature reactor, the supercritical water-cooled reactor, the molten salt reactor, the gas-cooled fast reactor, the sodium-cooled fast reactor, and the lead-cooled fast reactor.

TERRAPOWER TP-1 AND THE TRAVELING WAVE REACTOR

This is an impressive list of future reactor designs, but it leaves out one of the most forward-thinking concepts in nuclear engineering. This design, the traveling wave reactor, meets all the goals of the Generation IV initiative, but goes beyond these requirements, promising an unusually long period between refueling operations. It is being designed using the advanced methods that are critical to the Generation IV developments, and it may be added to the list of GIF reactors in the future.

The *traveling wave reactor (TWR)* is a variation of the sodium-cooled fast reactor design, which has been studied, prototyped, and tested for the past 60 years. To highlight the difference between the TWR and the typical breeder reactor, the traditional fast reactor breeder concept will be described first. In a traditional breeder reactor, the fission is sustained using fast neutrons, as are released by fission, as opposed to the moderated or slowed down neutrons used in a water-cooled reactor. These fast neutrons cause fission and a portion of heat transfer to the power generation function, but they are also used to transmute inert uranium-238 into fissile plutonium-239. The breeder reactor is refueled periodically, with new plutonium removed from the breeding blanket, formed into fuel assemblies, and used to replace the now burned-out fuel in the reactor center. A new charge of uranium-238 is loaded into the breeding blanket covering the reactor core, and the power production is restarted. The expended fuel is treated as radioactive waste material. In the breeder or any other conventional reactor, the level of neutron flux across the reactor core, from side to side, can be expressed as a symmetrical bump, diminishing to zero outside the core and rising to a peak value in the center. The level of fission and neutron activity always peaks at the center of the fuel-loaded core. The neutron level plotted against the width of the reactor is a classical standing wave, like an ocean wave that stands perfectly still.

The TWR has a larger than usual core, with 90 percent of its volume loaded with uranium-238, or depleted uranium. At the bottom of the core is a thin layer of fissile fuel, enriched with enough uranium-235 to begin a sustained fast-fission reaction. This starter fuel is enriched to about 12 percent uranium-235, or about three times the enrichment required by a conventional power reactor. The reactor is started, self-sustaining fission is established, and the reactor begins generating heat. The heat is transferred from the reactor core to a steam generator using liquefied sodium metal as the coolant, and the electrical generator is turned by the steam. The liquid metal coolant provides no neutron moderation, allowing excess neutrons to escape the fissioning portion of the core at high speed. The neutron flux profile across the vertical axis of the reactor core is stationary, but skewed. The peak neutron activity measured vertically is not in the center of the fuel mass, but down at the bottom, in the layer of fissioning fuel.

The uranium-235 at the bottom of the core eventually burns out, but the fission process does not stop. The uranium-238 next to the fuel at the bottom, incapable of fissioning in a productive way, has changed into plutonium-239, which is a legitimate reactor fuel. The adjacent plutonium starts to fission, taking up the power-production function. The burned-out fuel in the bottom stops fissioning. The peak of the neutron flux in the core has thus moved slightly, from the now destroyed uranium-235 to the new fissioning region in the core. The wave is now traveling, slowly, up the vertical length of the core.

Four fuel zones develop in the reactor core after the power has been produced for some time:

- ❋ **The depleted zone.** This region contains fission products and a subcritical fuel remnant. Fast neutrons from the fission process leaking into this zone can affect the fission products by activating them into rapidly decaying isotopes, reducing the activity of the waste products.
- ❋ **The fission zone.** In this region, the density of fissile material is sufficient to sustain a critical reaction. Only fast neutron fission is possible, as there is no moderator, so the fissioning fuel must be dense and fairly pure. Power is generated in this zone, and the neutron flux is at the peak value in the center.
- ❋ **The breeding zone.** This region is in range of the excess fast neutrons escaping from the fission zone, and the depleted uranium is being converted into fissile fuel. The density of fissile material has yet to reach a density that will support power generation.
- ❋ **The fresh zone.** The uranium-238 in this region is too far away from the fission to be affected by the fast neutrons, which are being stopped in the breeding zone. This region, which starts

out as 90 percent of the core, will be brought into the action when the wave reaches it.

Over time, the neutron activity peak, or the wave, travels from the bottom of the fuel load up to the top. When the top fuel has burned out, the reactor can no longer generate significant power, and it shuts down.

There is no lack of depleted uranium for loading TWR cores. Stockpiles in the United States, left over from uranium enrichment operations, amount to 772,000 tons (700,000 metric tons), which would produce $100 trillion worth of electricity. This conservative estimate assumes that no further uranium will ever be mined from the ground. Global stockpiles of depleted uranium are sufficient to supply the electricity demands of 80 percent of the world's population, at current per capita electricity demand levels in the United States, for 1,000 years. TWRs can also, in principle, burn depleted fuel left over from all power reactors now operating, as this fuel is primarily uranium-238, with small percentages of fissile uranium-235 and plutonium-239. Less than 1 percent of spent reactor fuel is fission product waste, and a TWR can achieve criticality with these wastes in place, once the deleted uranium has been up-converted to plutonium. Unlike the traditional breeder reactor, the TWR accomplishes fuel reprocessing on the fly, without the need to stop the reaction, unload the reactor, and ship the breeder blanket away for chemical separation of the fissile elements and fabrication into fuel.

These are all good reasons to pursue the TWR, but its most compelling feature is the lifetime of one load of fuel. The reactor is expected to produce power continuously for as long as 100 years on a single batch of uranium. There are no power outages for fuel swaps. The TWR has no need to be periodically cooled down and then dismantled, as is the case of every other reactor, in use or being designed.

The traveling wave reactor is not a new concept. It was first proposed in the Soviet Union in the 1950s, when it was called a breed and burn reactor. In the Soviet-style TWR concept, the neutron wave started on one side of the cylindrical core and proceeded to travel by circling around, ending up back at the burnt-out starter fuel. Further research in the United States was published in 1979 and 1988. In 1996, Edward Teller and Lowell Wood of the Lawrence-Livermore National Laboratory published the basis for the current model of the TWR. Various TWR core options have been explored ranging from cylindrical core and wave traveling vertically from the bottom to the top to rectangular cores and the wave traveling across

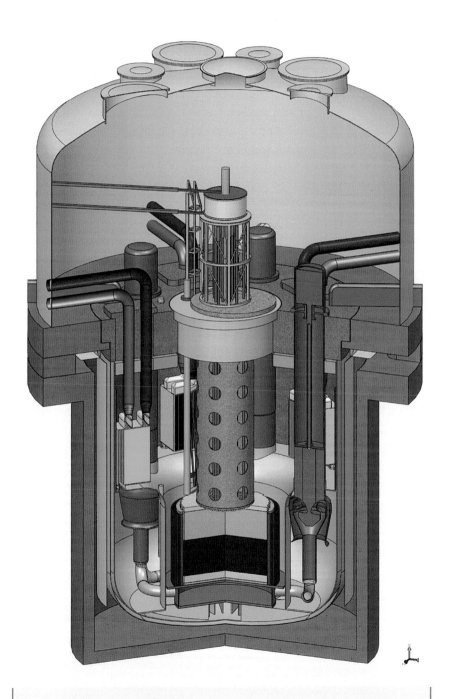

A cutaway picture of a TWR; the red part at the center is the unique core.
(Ash Odedra, TerraPower, LLC © 2010)

the core from one side to another. In the latest TWR version, the fuel is a metal alloy, formed into rods and loaded vertically in a cylindrical core configuration. The enriched starter section is in the center of the cylinder and the wave is kept standing at fixed position by occasional movement of fuel into the wave, like on a treadmill.

The TWR development project, started in 2006, was initiated by former Microsoft executive Nathan Myhrvold's company, Intellectual Ventures. Intellectual Ventures spun out a private company called TerraPower to pursue the idea with significant investment from Bill Gates (1955–), chairman of Microsoft. The plan is to have an operational test reactor, the TP-1, in 2020. TerraPower is headquartered in Bellevue, Washington.

The design of this reactor at TerraPower depends heavily on supercomputing simulations of its unusual core configuration and the advanced computer cluster built at Intellectual Ventures. All Generation IV reactors are being modeled to great precision using computers. Mathematical and digitally computed reactor models have been used since the earliest reactors built in the 1940s. Nuclear reactions to neutrons are always consistent and mathematically well behaved, and extremely accurate predictions of a reactor's behavior can be made. The precision of these predictions is dependent on the available computing power. The TerraPower team is using an interconnected array of 1,024 Xeon core processors assembled on 128 blade servers, currently under expansion to 5,280 Xeon core processors. This cluster is more than 5,000 times faster and more powerful than a desktop computer, and with this scale of processing the behavior of a virtual model of the TWR core can be studied in three spatial dimensions over its entire 100-year expected lifetime. Without the expense and licensing delays of building a reactor, the design can be tested, evaluated, and modified experimentally using these mathematical models programmed into a large computer array. The high-fidelity results of this type of modeling can ensure success for this and other bold ventures in nuclear engineering.

VERY HIGH TEMPERATURE REACTORS

The *very high temperature reactor (VHTR)* is interesting in that it is designed to operate with an unusually elevated coolant temperature of 1,832°F (1,000°C). Such a temperature can certainly be used to boil water into steam and drive a turbine, but this Generation IV reactor is intended for multiple purposes. In addition to electrical power production, it can also produce superheated process steam, which is used in a number of industries, from paper-making to oil refining. Its most productive function may be separating hydrogen from water molecules using high-temperature electrolysis. The production of bulk hydrogen may become an important factor in pollution-free personal transportation not relying on hydrocarbon fuel. (This topic will be fully explored in chapter 7.)

The VHTR is a thermal reactor, using moderated neutrons in the fission process. Because of the extremely high temperature in the core, graphite is used as the moderator and helium gas is the coolant. In a common pressurized water reactor (PWR), the highest temperature in the system, at the water outlet of the core, is about 600°F (315°C). There are two core designs under consideration, one using prismatic graphite blocks to make the moderator and one using fuel pebbles, made of a combination of uranium oxide and graphite. There is historical experience with both graphite concepts, but the pebble bed design may be favored. A prototype 10-megawatt pebble bed reactor, the HTR-10, is operational at Tsinghua University in China, where two full-scale pebble bed reactors are currently under construction. These projects should yield valuable insights into the economical fabrication and the long-term use of graphite-uranium pebbles.

The VHTR is an upgrade of the older *high temperature gas reactor (HTGR),* as first proposed by Farrington Daniels (1889–1972) at the Oak Ridge National Laboratory in 1947. This pebble bed concept, in which the reactor core is composed of spherical, graphite-uranium pellets, was seen as a workable structure with some benefits. A pebble bed reactor has no metal structure to support the fuel, so there is nothing to melt and collapse in an over-temperature accident. The reactor is built like a jar full of marbles, and the coolant is able to move through the core using the spacings among the randomly packed pebbles. The pebble bed core can withstand an over-temperature as high as 2,900°F (1,600°C) without damage, but this temperature would completely melt a metallic core structure.

The first prototype high-temperature pebble bed reactor was the *Arbeitsgemeinschaft Versuchsreaktor (AVR),* built at the Jülich Research Centre in North Rhine-Westphalia, Germany, in 1960. It was connected to the West German power grid in 1967 and produced 15 megawatts of electricity. Operation ceased in 1988. Much was learned from this experiment, but the exercise was not entirely successful. The site of the AVR is now considered to be the world's most contaminated nuclear installation, with about 200 fuel pebbles stuck in a wide crack in the bottom of the neutron reflector. Even if a reactor cannot be melted, high temperature still imparts a great deal of stress on the structure. Careful decontamination of the site will be ongoing for the next 60 years. Still, China copied its design to build the HTR-10.

The engineering challenge in the design of the VHTR is the fuel pebbles, whether they are piled loosely in the reactor vessel or held in ordered stacks in blocks of graphite. Fuel pebbles are typically the size of tennis balls, made primarily of pyrolytic carbon, a material very similar to graphite. A current design is the *tristructural-isotropic (TRISO)* pebble. The middle of the pebble is composed of uranium oxide, uranium carbide, or uranium carbonate, with a fissile component based on the uranium enrichment level. The sphere of fuel is covered by a porous layer of carbon, acting as both a neutron moderator and an absorber layer for gaseous fission products, followed by a dense layer of pyrolytic carbon. Over that is a very hard layer of ceramic silicon carbide to seal in the fission products,

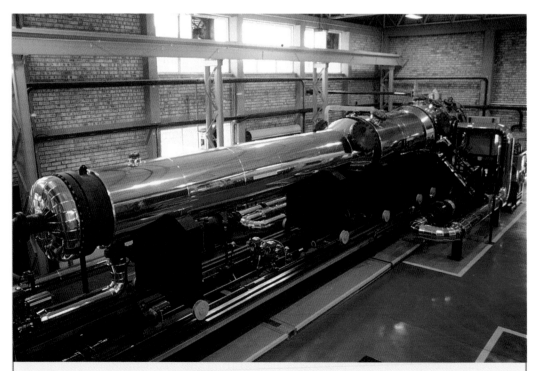

An experimental pebble bed reactor at the mechanical engineering department of Potchefstroom Campus of the North-West University in South Africa. The photo was taken in 2003. *(Graeme Williams/Newscom)*

and finally the ceramic layer is protected against breakage by another layer of tough pyrolitic carbon.

An advanced form of the TRISO fuel is the *QUADRISO* pellet, developed at the Argonne National Laboratory in Illinois. In this design, a layer of europium oxide or erbium oxide is deposited between the fuel and the first carbon layer. The europium or erbium acts as a burnable neutron poison. As the uranium-235 in the fuel fissions away, the reactivity of the reactor falls until it can no longer produce power. However, as the neutron poison is transmuted into nonpoison by neutron absorption, the reactivity available from the remaining uranium-235 increases. This allows a greater enrichment level for the manufactured fuel without increasing the fuel's reactivity to the point where it cannot be controlled. With the right combination of enrichment and initial poison, the reactivity of the fuel can remain almost constant over the significantly increased life of the core. A design goal of the Generation IV reactors is to increase the time between refueling operations, and the QUADRISO fuel could provide this feature.

There are problems with the VHTR still to be solved. There is nowhere in the pebble bed to place equipment for the usual reactor operation measurements, such as core temperature mapping and neutron activity levels. When producing power, it is good to have a complete picture of how heat is being produced on a moment-to-moment basis, always being alert to hot spots, but this will be difficult to achieve. The pebble bed is a black box, with its internal activity only estimable from outside measurements. Friction between pebbles tends to create dust, and this debris may act as a mobile fission-product carrier, transporting radioactive material around the cooling loop through the helium gas. Design effort is underway to work out these snags and make the VHTR practical for commercial power production.

SUPERCRITICAL WATER REACTORS

Most nuclear power reactors currently in use are light water reactors, or reactors using ordinary water as both a coolant and a neutron moderator. In the PWR, water is kept liquid by high pressure. Steam is generated by a second loop of water, passing through a heat exchanger with the super-heated water. In a boiling water reactor (BWR), water is allowed to boil in the reactor, becoming steam. These designs are complicated in different ways. The PWR requires two cooling loops, one for the reactor and one to run the turbogenerator. The BWR requires a steam separator at the top of the reactor vessel and jet pumps to circulate the bubbling water in the core. There is a third type of light water reactor. It requires no steam separator or jet pumps, and there is just one simple cooling loop that leads from the top of the reactor vessel to the turbogenerator and back to the bottom of the reactor through a water condenser. It is the *supercritical water reactor (SCWR)*.

Above a certain pressure, water cannot boil, regardless of its temperature. Liquid water cannot run a turbine, which requires steam, or water in the gaseous state. Steam can be made directly in a reactor, such as a BWR, but there must be liquid water covering the core. Steam cannot cool a reactor core. An alternate state of water that can both cool a reactor and run a turbine is supercritical water. At the extremely high pressure of 3,191 pounds per square inch (22 MPa) and a temperature of 705°F (375°C), water becomes supercritical. In this state, liquid water and steam are the same density. They are, in fact, indistinguishable. The same material can therefore be used for the two applications, a coolant/moderator and a turbine driver. In this unique situation, there is no need for steam separators in the vessel. No jet pumps are needed to circulate the bubbles, because there are no bubbles. There are no complicated, expensive steam generators, because the supercritical water is steam. With steam coming directly out of the reactor vessel, there is no need for a double cooling loop. The supercritical power reactor is simple and economical. Control rods in the SCWR are inserted through the top of the reactor, as is presently the configuration of common PWRs.

The research and development load necessary to make the SCWR a commercial power source will be concentrated on new materials for the reactor vessel and pipes. Although there is useful experience with supercritical water in fossil fuel power plants, the nuclear solution involves all the very high pressure and temperature with the added stress of high neutron radiation. Radiolysis of the water, breaking it down into hydrogen and oxygen under gamma ray bombardment, may affect the chemistry of the water. The chemistry of supercritical water under radiation is not well understood. Hydrogen embrittlement could alter the strength and hardness of the reactor vessel walls. These complications have always existed in nuclear power systems, but they may be amplified under supercritical water conditions.

The development of new materials is the key to adoption of the SCWR. Its attractive simplicity makes it worth a research effort, but its thermal efficiency is its most important feature. A current BWR or PWR is no more than 33 percent efficient, but the efficiency of an SCWR is an enticing 45 percent. It promises to use less fuel to generate more power, and as the global demand for nuclear fuel increases this asset may be moved to the front.

MOLTEN SALT REACTORS AND THE THORIUM CYCLE

The *molten salt reactor (MSR)* may have more impact on nuclear power in the future than any other Generation IV design. The overall design is novel and has been tested with working models, but the great advantage of the MSR concept is that it does not use uranium as fuel. It uses thorium.

The element thorium was discovered in Norway in 1828 and named for Thor, the Norse god of thunder. It is slightly radioactive, decaying in an alpha emission with a half-life of 14.05 billion years. It is estimated to be three times more abundant in the Earth's crust than uranium, and it occurs in nature in only one isotope, thorium-232. It is not fissile. Its industrial use has been limited to a few applications, such as gaslight mantles and tungsten arc welding. The demand for thorium has been slight, so its distribution and quantity are not known precisely. The U.S. Geological Survey (USGS), Mineral Commodity Summaries (1996–2010), as shown in the following table, represents the best estimate.

ESTIMATE OF ECONOMICALLY AVAILABLE THORIUM RESERVES IN METRIC TONS	
Country	**Reserves**
United States	440,000
Australia	300,000
Brazil	16,000
Canada	100,000
India	290,000
Malaysia	4,500
South Africa	35,000
Other Countries	90,000
World Total	1,300,000

Note: The quantities listed in this table are rough estimates, accurate to no more than two significant figures. The sum has therefore been rounded off to two significant figures, matching the level of accuracy in the data.

While no isotope of thorium is fissile, the available isotope of thorium readily activates under neutron bombardment into protactinium-233. The protactinium beta decays rapidly with a 27-day half-life into uranium-233, which is a fissile reactor fuel. This odd characteristic of thorium can be used to great advantage in nuclear power applications. Unlike mined uranium, in which only 0.72 percent of the refined metal is fissile, all of the thorium is usable. There is no need for expensive isotope separation and no need to bury more than 99 percent of the fuel as part of the fission waste. As a nuclear fuel, thorium is comparatively cheap, at about $30 per kilogram, whereas the cost of uranium has risen to about $100 per kilogram. The fact that most of the uranium is unusable makes thorium more than 300 times less expensive as a reactor fuel without considering the extra cost of enrichment.

Uranium-233 made from thorium-232 in a specially designed breeder reactor can be used immediately as nuclear fuel, just like uranium-235. A change in the reactor start-up protocol makes it possible to directly load thorium into a reactor core and use it as fuel. A starter mass of uranium-233 or uranium-235 is used to give immediate fission response, but as the uranium fissions away, excess neutrons transmute the thorium into protactinium, which becomes fissile uranium. The thorium-fueled reactor has a breeding rate of about 109 percent, meaning that after a year running at full power, the core still contains a full load of uranium-233 fuel plus a 9 percent surplus, and this situation ends only if the thorium is not replenished.

A molten salt reactor requires no fuel fabrication and no auxiliary breeding blanket to contain the nonfissile breeding stock. There are no fragile zirconium fuel assemblies and there is no fuel cladding. The fuel and the breeding stock are dissolved in the salt, which is also the coolant for the reactor core. The cooling system can contain far more than a critical mass of fuel. As the coolant-fuel mixture circulates in the cooling system, it becomes reactive only in the core vessel. The reactor tank is the only structure in the circuit that is shaped correctly for minimum neutron leakage and is sufficiently large to house a critical mass. Solid graphite, which is present only in the reactor core, acts as the neutron moderator to promote fission. In the circulation pump, the pipes, and the long, thin heat exchanger, the fissile material in the coolant cannot reach criticality. The heat exchanger transfers energy to a secondary loop using a molten salt containing no fissile material or radioactive fission products, and it drives a steam generator in a third loop using water. The water flashes to steam in this loop and drives the turbogenerator to make electrical power.

The primary coolant salt is a mixture of lithium fluoride, beryllium bi-fluoride, thorium tetra-fluoride, and uranium tetra-fluoride. The beryllium salt is a product of metallic beryllium in the reactor vessel reacting chemically with free fluorine, which can collect in the coolant when the uranium fissions into two atoms which do not make fluorine compounds. The beryllium scavenging of fluorine prevents corrosion in the cooling loop. The secondary salt is a mixture of sodium barium tetra-fluoride and

Looking down into the containment structure of the Molten Salt Reactor Experiment at Oak Ridge, Tennessee, in 1964 *(Oak Ridge National Laboratory, U.S. Department of Energy)*

sodium fluoride, formulated to run at a lower temperature than the primary salt. The primary salt is heated to 1,299°F (704°C) in the reactor, the secondary salt runs at 1,150°F (621°C) emerging from the heat exchanger, and the steam is superheated to 1,000°F (538°C).

RADIOISOTOPE THERMOELECTRIC GENERATORS

Nuclear power plants as discussed in this book are large costly installations, sprawling over several acres of land. They are unusually complex machines, having hundreds of remotely operated valves, dozens of pumps and fission control mechanisms, and multiple layers of emergency handling equipment. Miles of pipes and more miles of electrical wiring seem to knit the facility together, and constant vigilance and maintenance keep this complicated business producing power.

There is another type of nuclear power, on the opposite end of the spectrum of complexity, and it has been quietly built and developed over the past 60 years. This power source, the *radioisotope thermoelectric generator (RTG),* was first requested by President Dwight D. Eisenhower (1890–1969) in the early 1950s in his call for the development of peaceful applications for nuclear technology. The RTG has no moving parts and runs continuously with no human intervention or maintenance. It is scalable and can be made as small as a penny, developing a fraction of a watt, or it can be made as large as is required of the application.

The RTG uses the alpha decay of a radioactive isotope as a source of heat. The heat is applied to one side of a solid-state thermocouple, which converts the heat into direct current electricity. The idea of this thermal nuclear power source was considered long before nuclear fission was discovered, using the extremely energetic alpha particle emission of radium. Radium was discovered in 1898 and chemically isolated as a pure metal in 1910. Its energy release by nuclear decay was calculated to millions of times higher than any combustion process on a per atom basis, and speculative applications followed. One obvious use for radium was to drive a submarine underwater without the need for air. All it would take was several kilograms of radium immersed in a tank of water. The water would boil, making steam for a conventional steam engine. It was an interesting plan, but stocks of available radium amounted to micrograms and not kilograms. Radium eventually found practical uses illuminating watch dials and remote airfield landing lights.

During World War II, manufacturing processes for artificial heavy radioactive isotopes were developed, and one of these materials, plutonium-238, was of particu-

In designs of this molten salt power reactor, the fuel reprocessing cycle runs in a closed loop within the power plant. Fission products and extra fuel uranium-233 fuel made in the breeding process are chemically removed from the salt, and the thorium is replenished in a closed

lar interest for reinventing the plan for a radium-fueled engine. This isotope emits high-energy alpha particles similar to radium, and it can be produced in quantity by neutron capture in neptunium-237, which can be extracted from spent nuclear fuel. One ounce (28.4 grams) of plutonium-238 continuously produces 14 watts of power, and if left on a tabletop a pellet of it will start to glow red-hot. This energy can be used for direct heating, electricity generation by thermocouple, or both. The electricity production is only about 7 percent efficient, but the device has the great advantages of very rugged construction and having nothing that will wear out. After 87.8 years, the plutonium-238 will have slowly decayed to one-half of its heat-generating capacity.

In 1968, the Martin Marietta Corporation offered a portable generator using plutonium-238 for sale. Made of lead and shaped like a farm animal, it was named the pig, and guaranteed to operate underwater or in the most severe environments. RTGs found uses in remote arctic weather stations. These stations were not manned and had to operate for years without service or attention. The RTGs kept the equipment from freezing and provided the electrical power for a radio transmitter. A most appropriate application was found in deep space probes, for explorations beyond where solar panels can be used to generate electricity and where the temperature is close to absolute zero. The current American space missions Cassini-Huygens, New Horizons, Galileo probe, and Ulysses probe, among many others, are equipped with the general-purpose heat source radioisotope thermoelectric generators (GPHS-RTG), built at the Idaho National Laboratory by the Department of Energy. Each unit weighs about 165 pounds (75 kg) and produces 300 watts of electricity and 4,400 watts of heat. The fuel is 17.2 pounds (7.8 kg) of plutonium-238, and the electricity is produced by an array of silicon-germanium *unicouples.*

There may be a problem with full-scale implementation of this technology. Around 1990, with the fall of the Soviet Union, many remote RTG installations in Siberia were dropped from systematic surveillance. Hunters would find them in the wilderness, use them as sources of warmth in the Siberian winter, and die of radiation exposure. The RTG, for all its simplicity, is still a nuclear device, and it must be used with due care and diligence.

loop separate from the primary cooling loop. Power production using thorium produces lower levels of highly radioactive fission products than the uranium or plutonium fission process, because the fissile isotope is slightly lighter. The thorium process, however, is unique in that it produces protactinium-231. This isotope is dangerously radioactive, with a half-life of 31,000 years. It is removed from the coolant by chemical extraction in the reprocessing cycle and is disposed of as a waste product.

Extensive research into molten salt reactor technology began with the *Aircraft Nuclear Propulsion (ANP)* program in the 1950s, funded by the Department of Defense. At the Oak Ridge National Laboratory, a stationary model of a nuclear jet engine was built, developing 2.5 megawatts of heat. The reactor was meant to power a strategic bomber, so the power density had to be very high. Powering an airplane, the reactor did not have the usual luxury of unlimited weight and an ability to sprawl across a few acres of ground. For making the most power with the least amount of weight and size, molten salt was found to be an ideal coolant. The reactor could operate comfortably at 1,580°F (860°C), which was hot enough to heat the air in a jet engine. In 1954, the Aircraft Reactor Experiment (ARE), a nuclear reactor experiment designed for use in a bomber as an engine, ran at full power continuously for 1,000 hours. The ANP program was cancelled in 1960, before any nuclear jet engines were mounted in airplanes.

Further work with nuclear aircraft was not likely, but the molten salt concept had proven viable in the experimental work. The facility was immediately broken down and rebuilt as the *Molten Salt Reactor Experiment (MSRE)*. The new project was a 7.4-megawatt power reactor, built to test the idea of breeding uranium-233 from thorium. Construction was started in 1962 and completed in 1964. An engineering problem to be solved was the material used in the primary cooling circuit. The mixtures of salts to be tried all contained fluorine, and it is a terribly reactive element that quickly corrodes most industrial metals. A new alloy, Hastelloy-N, was formulated using chromium, nickel, and molybdenum, and all parts that touched the salt were fabricated from it.

Experiments with the molten salt reactor were performed successfully for five years, with the reactor running for the equivalent of 1.5 years at full power. In October 1968, the MSRE became the world's first reactor to operate on uranium-233 derived from thorium. In December 1969, the experimental program ended, and the MSRE was decommissioned in 2009. A proposed follow-on project to build a larger reactor at Oak Ridge

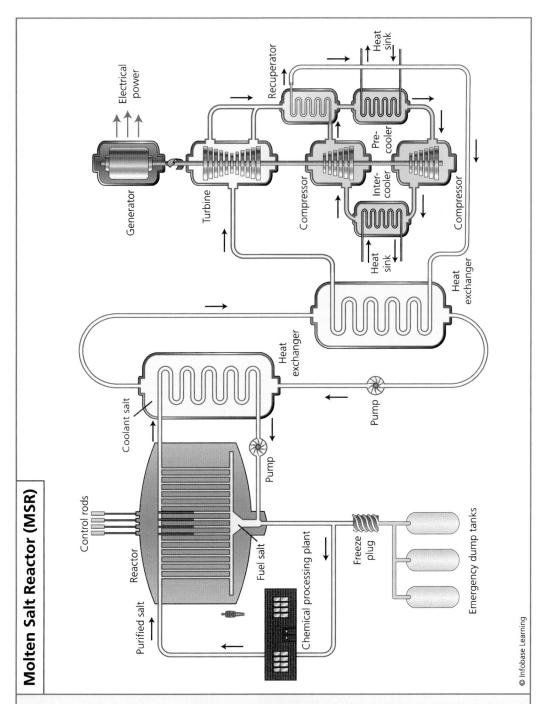

Molten Salt Reactor (MSR)

Generator

Electrical power

Turbine

Recuperator

Heat sink

Pre-cooler

Compressor

Inter-cooler

Compressor

Heat sink

Heat exchanger

Heat exchanger

Pump

Coolant salt

Pump

Control rods

Reactor

Fuel salt

Purified salt

Chemical processing plant

Freeze plug

Emergency dump tanks

© Infobase Learning

A diagram showing how a Generation IV molten salt reactor will work. It generates power using a primary cooling loop with fuel dissolved in it. Fuel is reprocessed on-site. A second molten salt loop with no dissolved fuel is used to boil water and make power.

in the 1970s using the results of the experimental program was not completed, and the MSRE was the last molten salt reactor built in the United States.

India, having a large reserve of thorium-bearing minerals and an increasing need for electrical power, has developed a thorium fuel cycle, extracting uranium-233 from neutron-irradiated thorium. The reactor to burn this fuel is a modification of the venerable CANDU design. The *advanced heavy water reactor (AHWR)* being developed at the *Bhabha Atomic Research Centre (BARC)* in Mumbai uses heavy water as the neutron moderator and pressurized light water as the primary coolant.

Using the MSRE work at Oak Ridge as a starting point, a company in Japan, International Thorium & Molten-Salt Technology Inc. (IThEMS), is working to develop a thorium breeder reactor. The full-sized reactor under design is the Fuji, but there are also plans to build a much smaller mini-Fuji with business partners from the Czech Republic.

In this century, as nuclear power occupies a growing percentage of the world's electrical power source and the demand for uranium increases, thorium mining, fuel reprocessing cycles, and reactors running on nothing but uranium-233 bred from thorium will become more important, possibly taking the lead in nuclear power development. In this case, old molten salt reactor designs from the 1950s will be reexamined and find new importance.

GAS-COOLED FAST REACTORS

An ability to breed new fuel from old, depleted fuel will become a very attractive feature of reactors in the future, so renewed attention is being paid to fast reactor designs. Fast reactors do not depend on a moderator to slow down the fission neutrons. The lack of moderator means that fast reactors produce fission at the less efficient high end of the energy scale, around 10 MeV. At this neutron speed, uranium-235 fission is inefficient, but alternate fuels work well. Thorium can be used, transmuting into fissile uranium-233. Plutonium-239 is a common fast reactor fuel, or both plutonium and thorium can be used in a mixed-fuel formula. Advantages of this scheme are that it does not rely on uranium, which may become costly, and the fuel reprocessing can be accomplished in a completely closed cycle, minimizing fission wastes. A breeding function for new fuel is also possible.

The old scheme of gas-cooling a reactor, used extensively by the British at the beginning of commercial nuclear power, is being reexamined as an efficient way to build a fast reactor. The proposed coolant is helium or carbon dioxide gas, to be used in a direct energy-conversion cycle, using no secondary cooling loops or steam generators. The gas can be heated to

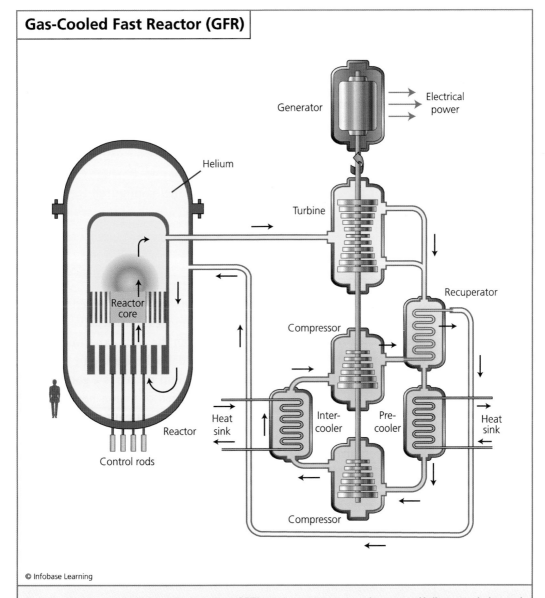

Gas-Cooled Fast Reactor (GFR)

© Infobase Learning

The Generation IV gas-cooled fast reactor (GFR) uses no steam to make power. Helium gas is heated to high temperature in the reactor core. It expands, and the hot gas is used to drive a turbogenerator directly. The helium is then cooled and compressed in two stages, preheated in its compressed state, and sent back into the reactor in a very efficient scheme.

1,562°F (850°C) in the reactor core and piped directly to a Brayton cycle gas turbine, which turns the generator and makes electricity. The gas is reconditioned for insertion back into the reactor by two stages of turbo-compressor. A conventional cooling tower, using water as the heat-transfer medium, is used to cool the gas in two heat exchangers, a precooler before the compressor stages and an intercooler between the turbo-compressors.

Proposed *gas-cooled fast reactor (GFR)* designs are small, producing only 288 megawatts of electrical power, but the use of economical fuel and the direct application of the hot gas make them attractive. The fuel configuration is an open design, and it is possible to fabricate it as blocks of material, plates, pebbles, or the traditional long tubes filled with cylindrical pellets. Design challenges include new construction materials for use in high-temperature conditions and safety studies for using the gas coolant. Ongoing demonstrations of this technology are the high temperature test reactor (HTTR) in Japan, which uses fuel configured in prismatic blocks, and the HTR-10 at Tsinghua University in China, which uses a pebble bed reactor.

SODIUM-COOLED FAST REACTORS

Sodium cooling is another old technology that is being retooled for use in Generation IV reactor designs. The advantages and disadvantages of sodium-cooled reactors are still in place from the early experiments with the experimental breeder reactor series in Idaho, beginning in the 1950s. Sodium or a mixture of sodium and potassium metals is a good reactor coolant because it does not moderate the speed of fission neutrons, and this is important for reactors burning and breeding plutonium. The heat conductivity of liquid metal is good, and the reactor can be run at a very high temperature without boiling the coolant, which leads to improved

(opposite page) A schematic diagram showing how the Generation IV sodium-cooled fast reactor operates. There are two liquid sodium coolant loops, operating in series, to make steam in the second heat exchanger and run the turbogenerator. The fast reactor core is disc-shaped, like a hockey puck. The inefficient core design wastes neutrons, which can be used to breed Pu-239 from U-238.

Sodium-Cooled Fast Reactor (SFR)

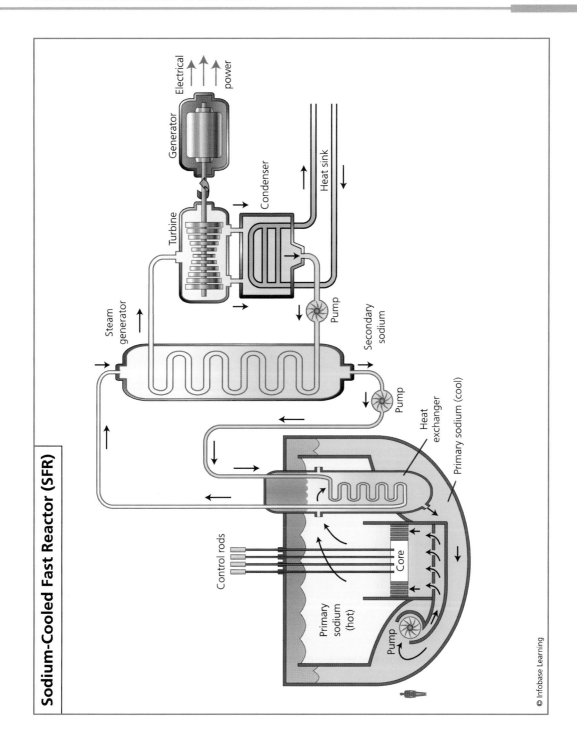

thermal efficiency when turning a turbogenerator. There is no chance that the coolant will cause a steam explosion. A major disadvantage is that liquid metal coolant reacts strongly with air or water. Unlike water coolant, a small leak is a potential disaster.

Arguments for the nonexplosive characteristics of a sodium coolant loop are compelling, and the breeding capabilities of the fast reactors are also attractive. However, steam explosions in water-cooled reactors have been rare and unusual, while many liquid metal reactors have been shut down simply because of coolant leaks. Fast reactors, such as the EBR-1 in Idaho and the Fermi-1 in Michigan, have seemed sensitive to slight modifications of the core configuration due to cooling problems. The sodium-cooled fast reactors may be successfully developed in the Generation IV program in spite of the history of problems with liquid metals based on a unique ability. The fast reactors tend to burn off, by fast neutron capture, the heaviest nuclear waste components. Plutonium, which builds up and must be disposed of in a water-cooled reactor, burns up as fuel in a fast reactor, and the fission products stabilize to lower levels of radioactivity in hundreds of years, rather than tens of thousands of years for the usual reactor waste. A water-cooled reactor uses only about 1 percent of the fissionable material in a fuel load, with the rest of it buried away, while a fast reactor burns almost everything.

The current Generation IV *sodium-cooled fast reactor (SFR)* is a combination of the traditional liquid metal fast breeder reactor designs and the integral fast reactor (IFR), a Department of Energy project that was cancelled in 1994, three years before completion. The IFR design was unique in its on-site fuel reprocessing and fabrication, using innovative electro-refining. In this process, the fission products are separated from used fuel using an electroplating method. The cleansed plutonium is then injection-cast into new fuel and put back in the reactor.

The refined SFR design uses double sodium cooling loops. The fast reactor core is immersed in a vat of liquid sodium and an electrically driven pump in the vat circulates it through the hot fuel. Heat is extracted from the molten metal by a heat exchanger immersed in the vat with the reactor, and the working fluid is also liquefied sodium. This hot sodium is pumped through a conventional steam generator, and the steam runs a turbogenerator in the usual way, with a water-loaded cooling tower acting as the ultimate heat sink for the power plant. The sodium coolant leaves the vat at 1,022°F (550°C). The SFR power station can be a small plant, generating 150 megawatts of electricity, or as large as 1,500 megawatts.

Prototyping the first SFR will be an expensive project and may require a multinational cooperative effort, but the rewards of efficient use of a boundless fuel supply may overcome all budgets.

LEAD-COOLED FAST REACTORS

The final entry in the competition for the Generation IV reactor design is the *lead-cooled fast reactor (LFR)*. This is an interesting variation on the liquid sodium reactor, in that the lead atom is much heavier than the sodium atom, and it imparts even less moderation on fission neutrons. This makes the LFR neutron energy spectrum extremely high, and this could lead to more efficient fuel breeding using uranium-238.

The proposed LFR runs at unusually high temperature, using nothing but natural convection to move the molten metal through the reactor core. Multiple heat exchangers are built into the top of the theoretical LFR reactor vessel, transferring hot gas to a Brayton cycle turbine in a closed external loop. The reactor core, using internal breeding, is expected to last 10 to 30 years between refuelings. No on-site fuel reprocessing is planned.

Although it sounds counterintuitive, the first experiments with lead as a reactor coolant were for nuclear-powered aircraft engines in the Soviet Union in the 1950s. The power density in a lead-cooled reactor is high, and a maximum power output could be achieved with a reactor of minimum weight, even with lead used as the working fluid. Lead-cooled engines were never used in an aircraft, but they did find use in Soviet Alfa class submarines in the 1970s. The OK-550 and BM-40A nuclear fission reactors produced 155 megawatts of thermal energy using a mixture of lead and bismuth as the coolant. Not only were the LFR engines lighter than water-cooled reactors of the same power output, they were also quieter. Without the gurgling sound of water being rapidly force-pumped through the core, a BM-40A was a nearly silent submarine power source.

The outlet temperature from the heat exchangers in the preliminary Generation IV LFR design is remarkably high, in the 1,380–1,470°F (750–800°C) range. If it were a bit higher, 1,530°F (830°C), then the LFR could be used to break water into hydrogen and oxygen, and this would be a direct and very economical way to make hydrogen gas, a road fuel of the

Lead-Cooled Fast Reactor (LFR)

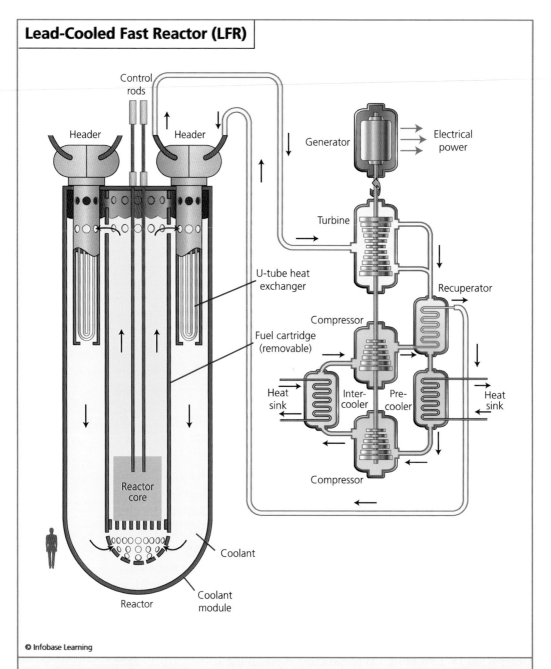

© Infobase Learning

The LFR uses lead as a coolant running at very high temperature. Heat exchangers located in the reactor vessel are used to transfer energy to helium gas, which drives a turbogenerator directly. The efficient compressor and cooler setup for returning the helium to the heat exchangers is similar to that used in the GFR.

future. Current designs are for small, modular LFRs, in the 50-megawatt range, and the use of these small power plants may be an economical way to expand nuclear power. The topic of modular power plants is discussed further in the next chapter, and the concept of using hydrogen to replace gasoline is covered in chapter 7.

4 Small Modular Reactors

For the past 50 years, the trend in nuclear power plant development and implementation has been to build power plants as large as is practical. This trend began soon after 1957, when the Shippingport Atomic Power Station began producing electricity and selling it to consumers. The Shippingport reactor produced only 60 megawatts of electrical power. A power plant, which was complicated and expensive to build, could be made twice as large, producing twice as much power and bringing in twice as much revenue, without spending twice as much money on construction. A 120-megawatt station was actually no more complicated than a 60-megawatt station, and it would fit on the same plot of ground bought for the smaller power plant. When the nuclear power construction rush began in the 1960s, it became apparent that the major cost of a nuclear reactor was not the special equipment that was required but the building site. The site had to be adjacent to a river, away from dense population areas, and large enough to accommodate cooling towers and spent fuel storage. Years of safety analysis were necessary to ensure that the site was not subject to seismic activity or environmental impact. The most precious part of a nuclear plant was the ground it was built on.

By 1982, when the Shippingport operating license expired, nuclear plants were in place producing 1,000 megawatts per reactor, with multiple reactors on one building site. This power level seemed a practical cutoff point. To make reactors much more powerful would require an advanced development of materials and cooling strategies. A one-gigawatt plant could be built using the same materials and coolant used for the old 60-megawatt reactor, and the business of making power was successful at this level.

The trend toward larger nuclear power plants makes sense for making a supply of bulk power reliable and available 24 hours a day. However, it has bypassed another mode of power generation with other advantages. Early in the development of nuclear power, there was also a push to make small, compact nuclear power plants. They were to be simpler than the big plants, easier to operate, and require less attention to maintenance. A small reactor, producing less than 50 megawatts of power, would put less stress on components, produce less fission waste, and burn up fuel so slowly it would not have to be refueled. It could be drop-shipped to a remote area needing electricity but having no fuel to burn in a conventional steam plant.

Research and development toward this type of reactor was conducted by the *Army Nuclear Power Program (ANPP),* beginning in 1954. There were special electrical power needs in the Antarctic and Greenland for remote radar stations and habitable research installations, under extreme cold without access to burnable fuel. A test reactor, the SM-1 (stationary, medium power, version 1), was built at Fort Belvoir, Virginia, in 1957. It produced 2 megawatts of electricity and was used as a training facility for army reactor operators as the program expanded.

In 1960, the PM-2A (portable, medium power, version 2A) power plant was installed at Camp Century in northern Greenland. Producing 2 megawatts of electricity plus space heating, the plant was shipped to the site in easily assembled pieces and was appreciated by the hundreds of people working in the extreme cold at Camp Century. In 1962, PM-1 was installed in Sundance, Wyoming, by the air force to power a radar station, and PM-3A went to McMurdo Station, Antarctica, for the navy. It provided 1.75 megawatts of electricity, heating of the buildings, and desalination of seawater for drinking and bathing.

In 1967, the first floating power reactor was built. The MH-1A (mobile, high power, version 1A) was assembled on a World War II Liberty ship, the *Sturgis*. It was towed to the Gatun Lake in the Panama Canal, where it floated while producing 10 megawatts of electricity and freshwater for the Canal Zone without disturbing the delicate rain forest environment of the region. A large-footprint power plant installation would have meant clearing land, shipping in coal, and dumping smoke into the air. In 1972, the ANPP was cancelled due to a lack of funding, and the MH-1A was shut down and shipped back to Virginia in 1977. The development and use of small reactors, easily installed and operated, ended as the general push for commercial nuclear power stalled.

As new needs for electrical power have emerged in the 21st century, there is a renewed interest in small-scale, modular power plants that can be installed without the enormous effort required for a one-gigawatt reactor. The needs for emissions-free power at a lower financial risk are growing, particularly in remote areas of habitation that are not connected to the larger power grid. It is simply not worth the cost to string long-distance electrical transmission wires to some locations, but people still need power in these places. There are presently several developments toward meeting these needs, and four of the most promising are described here. A sidebar describes the latest idea from the French nuclear industry. Although it seems the most radical in a group of bold designs, it could be the one that becomes reality.

THE NUSCALE REACTOR

NuScale Power in Corvallis, Oregon, is working on a small, modular nuclear power plant that generates 40 megawatts of electricity. Each module is a vertically mounted metal cylinder, 60 feet (18 m) tall and 14 feet (4 m) wide. The entire power plant is built in a factory, loaded on a tractor trailer or a train car, shipped to a location, installed, and hooked up to the local power-distribution network. A small operations staff is required, with most of the power plant functions being automated. The fuel is conventional power reactor fuel, as is used in most of the installed reactors worldwide. Uranium oxide pellets, 5 percent enriched uranium-235, are inserted into zirconium tubes and assembled into square bundles of 17 by 17 tubes, six feet (1.8 m) long. One fuel loading lasts for two years of continuous power generation.

It is a miniature pressurized water reactor using ordinary water for coolant, but it requires no coolant pumps. Hot, pressurized water around the reactor rises by convection to the top of the tall cylinder, where a small steam generator converts the heat energy to steam, and the cooled water then falls back down to the bottom to remove heat from the reactor. The steam is used to drive a small turbogenerator, and it is condensed back to water using an external cooling tower.

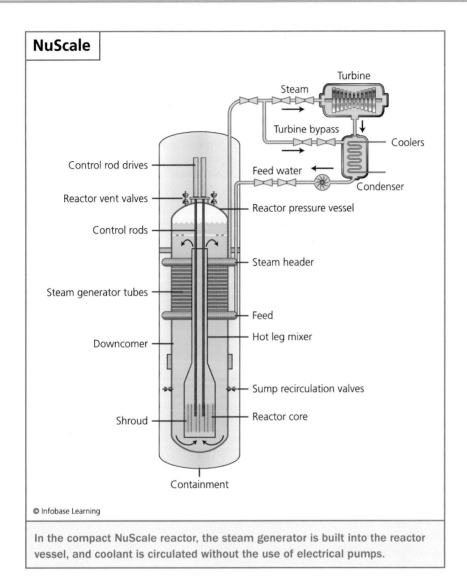

NuScale

Turbine

Steam

Turbine bypass

Coolers

Control rod drives

Feed water

Condenser

Reactor vent valves

Reactor pressure vessel

Control rods

Steam header

Steam generator tubes

Feed

Hot leg mixer

Downcomer

Sump recirculation valves

Shroud

Reactor core

Containment

© Infobase Learning

In the compact NuScale reactor, the steam generator is built into the reactor vessel, and coolant is circulated without the use of electrical pumps.

A NuScale generating facility can be scaled to whatever electrical power level is needed, in increments of 40 megawatts. Modules are simply stacked together, in a line, with up to 24 modules in one plant. A 24-unit NuScale plant would have almost the full power output of a single one-gigawatt reactor. Each reactor module is an individually sealed unit, with the radiation-containment capability of a large reactor's concrete, multilayer building. The steam pressure that stresses the reactor vessel and

Dr. José Reyes, standing in front of his NuScale reactor prototype *(Karl Maasdam Photography)*

pipes is half the level of a standard PWR, and all measures of safety are improved at this small scale of power generation.

The NuScale reactor design is based on the *multi-application small light water reactor (MASLWR)* developed by Oregon State University (OSU) and the Idaho National Laboratory starting in 2000. In November 2007, following tests on a one-third-scale model of the plant at OSU, NuScale Power was formed and the initial patents were filed. In February 2008, NuScale notified the Nuclear Regulatory Commission (NRC) of its intent to file a design certification request in 2012. The first plant is expected to be operational by 2018.

This modularized approach to nuclear power is addressing the problems of large-scale power production. Small countries in South and Central America, for example, that cannot necessarily afford or need one-gigawatt power stations, can easily afford to buy and operate these small, manufactured units. A full-scale plant can take a decade to build, but with this approach it can be bought from a catalog and shipped. Replacement parts, which have long been a problem with custom-built nuclear plants, are no longer a sticking point. The modules are standardized, and all use the same parts. A disadvantage to this approach, using standard reactor fuel for its economy, is the task of refueling. The fuel assemblies are half the length of full-sized reactor fuel, and they last half as long. Every two years, the reactor will have to be dismantled and refueled on site, using remote-handling cranes and equipment, and this may be more complication than the potential buyers are willing to accept. An even more attractive package would include a much simpler, off-site refueling procedure, in which the manufacturer takes care of fission waste disposal, and no special equipment is needed on-site for refueling.

Although the initial reactor design was funded by the Department of Energy, NuScale Power is privately owned and funded for detailed development of the technology. The Securities and Exchange Commission issued civil action against the principal investor on January 20, 2011, and NuScale must secure alternate funding to continue the final implementation of this design.

THE HYPERION POWER MODULE

Hyperion Power Generation Inc. is developing the same device, an inexpensive, modular nuclear reactor, from an entirely different engineering perspective. The power module is a stainless steel can, 8.2 feet (2.5 m) high and 4.9 feet (1.5 m) in diameter. It produces 70 megawatts of heat for desalinization or district heating or 25 megawatts of electricity if equipped with a turbogenerator. Each module is built in a factory and shipped by truck or rail to the site, where it is mounted in a below-ground concrete silo. One module contains sufficient fuel to run at full power for 10 years, and it is suggested that two units should be installed side by side, to ensure that one unit is producing power while the other has been sent back to the factory for refueling. The power module weighs 22 tons (20 mt). An aboveground metal building is included in the installation, housing the turbogenerator, steam condenser, condensate pump, and water tanks. Outside is located the cooling tower and a switch and transformer yard for the electricity transmission. Compared to a full-sized power plant, the Hyperion module should be inexpensive and very easy to erect. One reactor module, plus auxiliary equipment, will cost about $30 million, installed. It promises all the advantages of nuclear power, reliably producing power year after year without emitting any noxious gases. In September 2009, Hyperion projected a sale of 4,000 units in the United States, Britain, and Asia, and expected to ship its first unit in June 2013.

The design for this power reactor began at Los Alamos National Laboratory (LANL) in New Mexico in 2002. The scientists Otis Peterson and Robert Kimpland designed the Compact Self-regulating Transportable Reactor (ComStar), using uranium hydride as the fuel. This design had the safety characteristics of the TRIGA reactor, as described in chapter 2, in that it was self-controlling. The hydrogen component of the fuel would lose its moderating function if it ran too hot, and this would shut down the fission reaction until the hydrogen had cooled down to the correct operating temperature. This was seen as a good characteristic of a reactor that would have minimum operating supervision, and the design was licensed to a new company, Hyperion Power Generation in Santa Fe, New Mexico.

The uranium hydride design had good characteristics, but it would require a great deal of expensive development, and the company needed a design that could be put into production without decades of experimentation. In November 2009, a completely new design was announced. Instead of a thermal reactor using the hydride fuel and a hydrogen gas coolant, this one uses uranium nitride fuel, it uses fast fission instead of thermal fission, and the coolant is a mixture of lead and bismuth. This technology is very similar to that used in the Soviet Alfa Class submarines, as described at the end of chapter 3, and it may be easier to commercialize than the revolutionary uranium hydride design. This liquid metal coolant design has the advantage of not exploding and scattering fission products if it should overheat. Leaked coolant will solidify immediately and will not evaporate into the atmosphere or wander into the soil. A double loop of molten lead-bismuth ensures that a primary coolant leak cannot contaminate the steam-generating equipment or anything aboveground. The fast-fission reactor core contains enough excess fuel, slightly less than 20 percent enriched uranium-235, to run the reactor a long time. The limit of

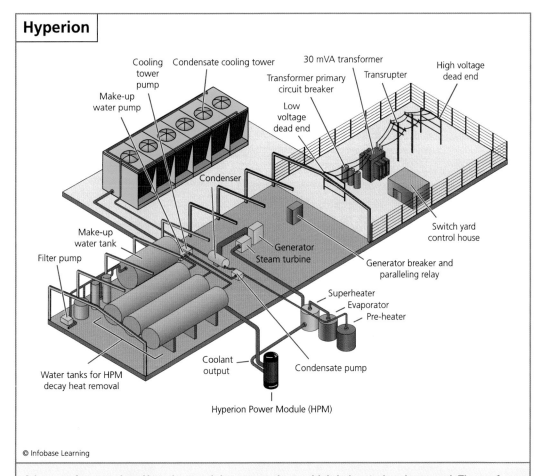

Hyperion

© Infobase Learning

A layout of a complete Hyperion modular power plant, which is located underground. The roof of the metal building that houses the turbogenerator and water tanks is cut away to show the internal parts of the system.

20 percent enrichment was set by the Nuclear Non-Proliferation Treaty of March 5, 1970, which defines any uranium enriched to 20 percent or more to be atomic bomb material.

Because it uses molten metal as its coolant, this reactor works at atmospheric pressure, and there is no explosive steam buildup in the reactor vessel. Quartz is used as the neutron reflector surrounding the core, and the control rods are made of boron carbide. These materials are unlikely to melt in an uncontrolled fission accident, and boron carbide pellets can be released to fall by gravity into the core and shut it down quickly in an emergency.

An additional incentive to buying this module is that Hyperion will remove the reactor from its silo, transport it back to the factory, refurbish and refuel it, return it to the site, and take care of disposal of the nuclear waste. This last service, the waste disposal, makes the Hyperion module

THE FLEXBLUE NUCLEAR POWER PLANT

On January 24, 2011, a French consortium consisting of the nuclear reactor vendor Areva, the world's largest utility company Électricité de France (EDF), the Commissariat à l'énergie atomique (CEA), and the defense company Direction des Constructions Navales (DCNS) launched a two-year feasibility study of a unique modular nuclear plant named Flexblue. It is to be a self-contained, sealed tube, about 330 feet (100 m) long and 50 feet (15 m) in diameter, with a complete nuclear power plant inside. Each Flexblue module will generate 50 to 250 megawatts of electricity. The modules will be completely waterproof and are to be sunk about a mile offshore. They will sit on the seabed at a depth between 200 feet and 330 feet (60 m and 100 m). Heavily insulated transmission lines will conduct the electricity back to shore, and a steel cable-mesh tent over each unit will protect it from ships' anchors.

These power plants will require no resident operators or maintenance personnel. What little control and fault analysis they will require will be by remote control, over electrical cables packaged with the power lines. They will be built with a Generation III safety level, and several can be stacked next to each other on the seabed, making a 1 billion watt power plant consisting of four or five modules.

The Flexblue is basically a French nuclear submarine without a propeller, crew's quarters, or onboard weapons systems. This means that the technology involved with very attractive. Remotely located mining and telecommunications companies in South Africa, the Czech Republic, and the United Kingdom have ordered 150 units.

A disadvantage to this design for an inexpensive modular reactor is that a working prototype has not been built. The NRC is aware of this development project but has yet to receive the request for design approval or a license to build it.

THE TOSHIBA 4S REACTOR

Toshiba and the Central Research Institute of Electric Power Industry of Japan are working on a modular reactor design similar to the Hyperion, although this project is much farther along. The Super Safe, Small and Simple reactor, the *4S,* is built into a sealed cylindrical vault 98 feet (30 m)

the design and construction of the Flexblue is well proven, having been used successfully and with a perfect safety record since the 1950s. The advantages to this approach are interesting. A foremost advantage is that there is no need for a cooling tower or proximity to a river. All of the secondary cooling is provided by the almost infinite heat sink of the ocean. Seawater is simply sucked in through a port and blown out another port at slightly higher temperature. A major accident, involving a catastrophic system failure with radiation release, would happen in an ideal setting. At the bottom of the ocean, radiation exposure to living things is prevented by the excellent shielding characteristics of water. Any gaseous or liquid released radioactivity is diluted to zero by the volume of the ocean. Theft of fuel or radioactive fission waste is prevented by the technical difficulty of surreptitiously breaching a submarine hull in deep water. For the initial startup and for maintenance operations, the power plant will be reached by minisub, and personnel will gain access though a special hatch. At the end of its usable life, it would not be unreasonable to simply leave this power plant in place, as its residual radioactivity decays away in the safety of an underwater resting place.

This nuclear plant concept bears watching. It may well be a credible path toward what many are trying, to build a marketable mini-reactor. The major question may be, how big is the market for this manufactured product?

underground, with a 36-foot (11-m) high service building aboveground on a foundation 72 feet by 52.5 feet (22 m by 16 m). As a power plant, the 4S produces 10 megawatts of electricity. A 50-megawatt version is in the planning stages.

Like the Hyperion, the 4S employs fast neutron fission and molten metal as a coolant, giving it a coolant outlet temperature of 950°F (510°C). The reactor core is assembled as 18 hexagonal fuel assemblies, loaded with a uranium-zirconium alloy. The enrichment is the same as the Hyperion, 19.9 percent, and the same three cooling loops are used. An interesting feature of the 4S is a mobile, ring-shaped neutron reflector, which slowly migrates upward on the reactor core, for more than 30 years, as the fuel burns. The core is long and thin, and the fuel fissions preferentially where the reflector ring sits. Toshiba filed a letter of intent for design certification with the NRC on March 23, 2010.

The Lawrence Livermore National Laboratory was contracted to do a design study of the 4S, and the results suggest that lead would make a better coolant, as used in the latest Hyperion design. Lead is more transparent to fast neutrons than sodium, and it is more opaque to gamma rays. It also does not react with air or water if it should leak out.

This finding is interesting, but the biggest obstacle to modular reactor implementation is not technical, nor is it safety related. Insurance and legal requirements are the last hurdle. Unfortunately, the obligatory burdens of liability insurance and regulatory compliance are written for full-scale power plants, generating more than 2 billion watts of electricity. These very small, modular power plants have a proportionally small fraction of the site impact, potential maximum hazard, and overall worth of a large installation, but the costs are the same. The NRC is aware of this problem and is working on alternate schemes that will make small reactors financially feasible.

Toshiba already has its first customer for a 4S reactor. The city of Galena, Alaska, is not accessible by road. With a population of about 612, it is the largest city in the Yukon-Koyukuk Census Area, and its energy needs are met using fuel oils, gasoline, and propane. The Yukon River is frozen solid eight to nine months of the year, so fuel for the winter must be air-shipped, and this makes it unusually expensive. In 2006, electricity in Galena cost $0.68 per kilowatt-hour.

On December 14, 2004, the Galena City Council voted to allow Toshiba to install and test its new 4S power plant. Installation will be free of charge, and the citizens of Galena will pay only the operating costs. The cost of energy in Galena is expected to fall to $0.05 per kilowatt-hour, and the 10 megawatts will be enough power for everyone. Plans are to install the 4S in Galena by 2013. Galena will be supplied with inexpensive energy, and Toshiba should gain favorable publicity and proof that its 4S reactor will operate in a cold, remote climate.

THE BABCOCK & WILCOX MPOWER REACTOR

The Babcock & Wilcox Company (B&W) of Charlotte, North Carolina, built the reactors at the Three Mile Island Nuclear Power Station in Pennsylvania. Unit 2 at Three Mile Island suffered an unfortunate core meltdown in 1979, and afterward B&W had trouble selling commercial power reactors. The company did, however, remain active in the nuclear power business, building special components for the nuclear navy and NASA. B&W has reemerged into the commercial nuclear market with a highly competitive product, a modular nuclear power plant called *mPower*.

The mPower is considered a Generation III++ design, more advanced than any Generation III+ reactor but more readily available than any Generation IV reactor. As seems the case with all modular reactor designs, the mPower will be installed in an underground silo. It will be manufactured in a facility owned by B&W, using steam-supply parts manufactured by B&W, and can be freighted to the power plant site by rail. Each unit can produce 125 megawatts of electricity, and it will not be difficult to install six of them at one site, making a full-sized, 750 megawatt power plant. Depending on local needs, from one to more than 10 units may be installed in one location. The reactor will run for 4.5 years on one fuel loading, and spent fuel can be stored in an on-site pool for 60 years.

This reactor is a miniaturized, simplified version of the standard pressurized water reactor (PWR) used in most of the world's nuclear power plants, and as such it should be easier to pass the NRC design review and licensing requirements than some of the more exotic designs. Similar in concept to the Toshiba 4S, the reactor core is at the bottom of a long, sealed metal tube. Without all the complicated external plumbing of a conventional power reactor, the steam generator is located in the same tube, at the top, and water circulates between the steam generator and the reactor core by natural, convective circulation. Water in the reactor core

is heated, and the hot water rises in a column above the reactor into the steam generator. Water from the turbine loop goes into and out from the steam generator, picking up the heat and turning the water into steam. The water at the top of the steam generator, now cooled by the water from the external loop, falls over the top of the generator and down the sides, back to the bottom of the reactor at the base of the vertical tube. Small turbo-pumps, called circulators, are built into the outer water path to encourage this action. This fully integrated primary cooling system greatly reduces the chances for pipe breaks and leaking valves. There is no pressurizer or steam-relief valve, the components that caused the Three Mile Island meltdown accident. If the system fails, the reactor is protected from melting temperature by the fact that all of the primary coolant is contained in the same vessel as the core. It can remain in safe condition indefinitely, with the entire electrical system shut down and no power to the water pumps.

Standard PWR reactor fuel and fuel assembly structures are used, so there is no need for a special, costly new fuel fabrication plant for these reactors. The reactor controls work like a common PWR, with the neutron absorbers introduced into the top of the core and moving up and down to keep the reactor precisely critical in all situations. Refueling is

A cutaway picture of the B&W mPower reactor. As is the case with some other modular reactor designs, the steam generator, shown in the top half of the picture, and the reactor core are both located in the same steel container. (© 2010 Babcock & Wilcox Nuclear Energy, Inc.)

simplified, as the entire core can be lifted out as one unit and replaced with a fresh one.

B&W may already have a first customer for its mPower. The Tennessee Valley Authority (TVA) has issued a letter of intent to the NRC to apply for a license to operate a cluster of six mPower units, to be located on the site of the cancelled Clinch River Breeder Reactor Project near Oak Ridge, Tennessee. The cost per unit will be about $500 million, and all six will be $3 billion, assuming that B&W will not give a quantity discount. This is still an expensive power plant, but the funding schedule for a modular plant is different from a full-scale single or double reactor station. TVA will first buy two units, install them, and collect enough revenue from them to then buy two more. The initial two will then begin paying for themselves, while the second two make money to buy units five and six. By staging the costs, the entire $3 billion is not tied up in one large reactor. As a final advantage to the modular design, when the plants are finally shut down after decades of use, the underground silos can simply be filled with dirt, with most of the reactor inside. The cost of the plant teardown is minimized.

In general, the plans for implementing modular reactors worldwide are good. Small, low-power units can be distributed more evenly in power grids, and the loss of one small power plant due to a mechanical breakdown is less of a disaster than the loss of a large plant, supplying a great deal of the base power to a region with a single machine. Building a nearly complete power plant in the controlled environment of a factory is far superior to erecting a large, complicated machine in the field from stacks of parts, shipped in from all over the world. It should cut down the time needed to implement a power plant, as well as the cost. Small, remote communities, such as the one in Galena, can have all the advantages of modern electrical power without the impossible task of erecting transmission lines. The waste of power lost by being shipped overland by high-voltage transmission lines is cut down, as the consumer can be closer to the source if very small plants are installed near communities. Small nuclear plants mean small disasters, if the worst should happen.

These small nuclear power sources may be an important component of power generation in the near future. On the opposite end of the scale of future nuclear installations is the subject of nuclear propulsion. It is an application of nuclear power for which there is no immediate need, and yet it may be something for which we should be planning. The universe is opening up to us as never before, with the newly improved telescope images of stars and galaxies, and Earthlike planets are now being discovered on a monthly basis. It may be time to start thinking about how physical probes and eventually mankind can travel to these very far destinations. This may be the far future, but the next chapter will detail the surprising steps that are underway in anticipation of it, and it all involves energy from the nucleus.

5 Nuclear Propulsion for Extended Distances

It is difficult to predict the future: Economic stresses or the global impacts of political changes can ruin the most careful predictions, throwing the trajectory of a carefully planned future off track. There is, however, a track of the future that cannot be derailed. If humans ever decide to get off this planet and go somewhere else, for exploration, resources, expansion, or just because it is there, then it will have to be accomplished using nuclear reactions. Chemically powered rocket engines, for all their power and utility for achieving Earth orbit or sending probes out among the planets, simply lack the energy density necessary for very long flights at high speed. It is possible to send a lightweight probe out of the solar system using chemical rockets, but it takes decades to make the trip, and there is a great deal of empty space between our solar system and one to be visited. Much faster, heavier spacecraft will be necessary when we begin to plan for deep space exploration. Possible targets are being discovered, studied, and cataloged now, and the next step of exploration will have to involve not chemical reactions, but nuclear reactions. The power available per unit mass or volume from a nuclear reaction is at least 1 million times better than that from a chemical reaction.

The burn time available from a set of tanks on a chemical rocket containing hundreds of thousands of gallons of fuel is measured in seconds. The burn time available from a nuclear propulsion engine is measured in years. That is the basic, important difference in these two means of escaping Earth, chemical and nuclear propulsion. This chapter will explore several imaginative designs, none of which have been tried. One or all of them could be in our distant future. The sidebar describes an actual planned mission and a propulsion system straight out of science fiction, where some important ideas have begun. Nuclear propulsion is a field in which anything is possible and engineering minds are allowed free rein.

FISSION FRAGMENT ROCKETS

The performance of rocket engines is expressed in terms of specific impulse, which is the improvement in momentum of the rocket vehicle per unit amount of propellant used, expressed as seconds, or as kilonewton seconds per kilogram in SI (International System of Units, abbreviated from the French Système international d'unités) notation. A solid rocket engine, such as those used as boosters for the now obsolete space shuttle, has a specific impulse of 250 seconds (2.5 kN·s/kg). The most efficient bipropellant liquid-fuel rockets, such as the liquid hydrogen/oxygen rockets used in the upper stages of the *Apollo* moon rocket, have a specific impulse of 450 seconds (4.4 kN·s/kg). The proposed fission fragment rocket has a specific impulse greater than 100,000 seconds (986 kN·s/kg).

This extraordinary rocket engine was designed by nuclear scientists at the Idaho National Engineering Laboratory and the Lawrence Livermore National Laboratory in 1988. The method of using nuclear fission to power a rocket is to use it as an energy-release mechanism, to heat fuel to a high temperature. The heated fuel then escapes rapidly out of the rocket nozzle, resulting in forward thrust—the higher the temperature, the higher the attainable speed for the rocket. The normal temperature of fission is millions of degrees, which is quite good for rocket propulsion, but this is only the temperature of individually fissioning nuclei and not the average temperature of the reactor. Individual fissions spread the energy, and thus the temperature, to adjacent, nonfissioning nuclei, and the temperature averages out to thousands of degrees at most, and not millions. If it were possible to single out nuclei as they fission and use these directly as a rocket fuel, then the extremely high temperature of fission could be put to good use. There would be no need to inefficiently heat up a separate fuel substance with the fission. The hot debris left over from a recent fission would become the rocket fuel.

This concept is possible. If the surface area of the reactor fuel can be maximized, to the point where nearly every atom of fissile fuel is exposed on a surface, then any fission debris will tend to blast off from the surface with its full 200 MeV of energy, not sharing any of it with neighboring nuclei which have not yet been hit with a fission-causing neutron. These fission fragments are naturally ionized, having no associated electrons turning them into neutral atoms, and as charged particles they can be directed at will using static magnetic fields. The extremely hot, energetic fragments can thus be focused into a beam, out the back of the vehicle, giving it extremely efficient thrust forward.

The extreme surface area is accomplished by plating the fissile fuel, uranium-235, onto long, thin carbon fibers, each having a tremendous surface-area-to-volume ratio. The fibers are then collected and arranged into a disc, with the fibers pointing out like spokes in a bicycle wheel. Four or more discs are then mounted on a common shaft, separated from each other by sufficient distance to make them subcritical. There is enough uranium-235 in the discs to make several critical masses, but the unchecked speed of the neutrons traveling from disc to disc does not encourage critical fission. Neutrons are released spontaneously on occasion, but they are fast and unmoderated and incapable of starting a runaway chain reaction.

Next to the discs is a cylinder of solid moderator material, such as graphite or beryllium. It has four slits cut in it, so that the discs can rotate through it smoothly. In the opposite side of the moderator cylinder is cut a slot, leading to an array of permanent magnets and the rocket nozzle. To start the engine, the discs are moved into the moderator slots. Given a neutron moderator between discs of uranium, thermal-neutron fission begins in earnest. It can be kept exactly critical by adjusting the depth of the discs into the moderator cylinder. The power level can be lowered or raised, by making the reactor temporarily subcritical or supercritical, and then returning to exact criticality by in and out motion of the common shaft. The portions of the discs that are in the moderator slots become an active power reactor.

Uranium-235 nuclei, capturing moderated neutrons emitted by fissions on adjacent discs, fission and blow off the surface of the fibers. The extremely hot fragments, heated to millions of degrees, are directed by a magnetic field to the nozzle, where they leave the spacecraft at high speed. What is left of the uranium underneath the fission has been heated to such a temperature that it can no longer support fission. The hot uranium atoms have expanded too far apart by being heated, and they must be cooled down before another series of fissions can occur.

Fission Fragment Rocket

Fissionable filaments · Revolving disks · Reactor core · Fragment's exhaust

© Infobase Learning

The fission fragment rocket. A very special fuel configuration in a nuclear reactor causes fission fragments to escape and cause thrust. Of all the advanced rocket propulsion methods that have been proposed, this may be the most practical.

To cool the fibers, they are slowly rotated out of the moderator as the common shaft turns. By the time a complete revolution of the discs is completed, the fibers have cooled down to a temperature that will once again support fission, and the process continues. There is a constant blast of fission products streaming out the back of the spacecraft, as new uranium surfaces are rotated into the moderator. As the fuel is burned up, the discs are inserted more deeply into the moderator to maintain criticality at the selected power level. The reaction shuts down after enough uranium-235 nuclei are gone that fission can no longer be supported or when the discs are withdrawn from the moderator cylinder. There is no radioactive fission product buildup, because it leaves the spacecraft through the engine nozzle into the infinite volume of space.

It is a good design, and there is no obvious reason why it cannot work. With its very high specific impulse, it is the type of rocket that will be necessary for preliminary interstellar missions. In 2005, an improved fission fragment rocket was proposed by NASA. It is called the dusty plasma bed reactor.

The only thing with more surface area than a thin fiber is a dust particle. Form the uranium-235 fuel into dust, or nanoparticles, and suspend them in a vacuum chamber, or the vacuum of outer space, using an active electromagnet. The electromagnet is a solenoid, surrounding the chamber containing the dust, and this combined with an external electrical charge acts as a bottle to contain the dust in suspension. The dust is excited to fission by the presence of a moderating neutron reflector, built of beryllium oxide or lithium hydride, surrounding the solenoid. Criticality is controlled by varying the amount of reflector that is affecting the fission process. Fission fragments blast off the surfaces of the dust particles, driven by the millions of degrees of heat generated by the fissions, and they escape preferentially through one end of the magnetic bottle, leaving at high speed through the rocket nozzle. Cooling of the remaining fuel back down to fission temperature is no problem. Because of the extremely small size of the fuel particles, they are unable to retain heat, and it radiates away into the vacuum chamber and contributes to the energetic exhaust.

The efficiency of this scheme is more than 90 percent, and it should be able to achieve a specific impulse of more than 1 million seconds (9,860 kN·s/kg). The exhaust velocity of this engine is 3 to 5 percent of the speed of light, or as high as 9,300 miles per second (1.5×10^7 m/s).

THE FUSION ROCKET

The efficiency, the specific impulse, and the exhaust velocity of these nuclear rocket engines are impressive and all far beyond what can be achieved with chemical rockets. However, another factor must be considered when planning a space mission. The thrust of the engine is also important. To lift a spacecraft weighing many tons off the ground requires a thrust that is greater than many tons, and chemical rockets are ideally suited for making enormous thrust values for short times. Nuclear rockets as described in this chapter produce thrust in ounces or grams and not tons. They can potentially run continuously for years and reach sub-light speeds without running out of fuel, but the acceleration is slight compared to a chemical rocket. A chemical rocket operates like a race car. It is powerful but wastes a lot of fuel. A nuclear rocket is like a hybrid car. It is not very powerful, but it uses every molecule of fuel to good effect.

If nuclear fission products can act as rocket exhaust, then a fusion reaction should also be usable as an engine. Fusion and fission are on opposite ends of the scale of nuclear reactions. In fusion, light nuclei are combined into one nucleus and energy is released. In fission, a heavy nucleus is split into two lighter nuclei and energy is released. If a fusion reaction can be initiated in a rocket engine, then the energetic ions resulting from the reaction, heated by fusion to millions of degrees, can be directed to leave through a nozzle using magnets. The concept of a fusion engine is thus very similar to that of a fission fragment rocket, with the major difference being the nature of the nuclear reaction.

Many fusion reactions release neutrons. Unlike fission reactions, in which neutrons are very useful for sustaining a continuous string of reactions, fusion neutrons are a waste of energy. Unlike the ions produced by fusion, the neutrons fly off in all directions and cannot be channeled by a magnetic field to leave through the nozzle, producing thrust. Any thrust produced by a quantity of accelerated neutrons sums to zero. Fusion for space propulsion must therefore be *aneutronic,* producing no particles without an electrical charge. A possible aneutronic fusion is the deuterium-helium-3 reaction:

$$D + {}^3He \rightarrow {}^4He + p + 18.3 \text{ MeV}$$

Both the helium-4 ion and the proton products of this reaction are charged and can be magnetically directed, so all of the 18.3 MeV of energy is usable as thrust. The deuterium component of the fuel can be extracted from seawater. Helium-3 is a rare commodity on Earth and will have to be manufactured using nuclear reactions, but it is believed to exist in exploitable concentrations on the lunar surface. Using the Moon as a jumping-off point and a source of fuel, interplanetary missions could be executed using fusion engines.

To attain fusion in deuterium and helium-3, these gases must be reduced to a plasma state, confined in a small space, heated to millions of degrees, and compressed to the point that nuclei are in close contact. Two ways to achieve this have been proposed: a form of continuous magnetic confinement and pulsed inertial confinement, or laser fusion. For a magnetic fusion rocket, NASA's Glenn Research Center at Lewis Field, Ohio, has proposed a new vehicle design, *Discovery II.*

The *Discovery II* is 787 feet (240 m) long and can carry a manned pay-load of 190 tons (172 metric tons) to Jupiter in 118 days or to Saturn in 212 days using 960 tons (870 metric tons) of helium-3 and deuterium propellant. Its engine is a small tokamak fusion reactor, made spherical. The exhaust velocity of the propellant, and therefore the top speed of the vehicle, is 216 to 288 miles per second (348–463 km/s).

Another NASA laboratory, the George C. Marshall Space Flight Center at Redstone Arsenal, Alabama, has proposed a smaller spacecraft for a similar mission, using a variation of laser fusion, called *magnetic target fusion (MTF)*. A NASA study the Human Outer Planet Exploration (HOPE) has designed a spacecraft that is 360 feet (110 m) long. It carries 117 to 182 tons (106–165 metric tons) of fuel and can deliver a 181-ton (164 metric ton) payload to Jupiter's moon Callisto in no more than 330 days.

It has a higher exhaust velocity than the *Discovery II*, 435 miles per second (700 km/s), but does not claim as much thrust. The engine heats the fuel mixture to plasma and confines it in a magnetic field, but unlike the tokamak it does not compress the plasma into a fusion state continuously. It pulses the fuel to compression using tightly directed radiation beams. Lasers have been used for this purpose in Earth-bound experiments, but they are too heavy and bulky for use in space. Instead, the HOPE space-craft engine uses particle accelerators to provide the energy beams.

There are two problems with implementing these ambitious plans for manned exploration to the outer planets. First, any fusion reactor using current technology is extremely large and heavy. The National Ignition Facility in Livermore, California, is the latest inertial confinement fusion experiment, and it is the size of an airplane factory. ITER, the current experiment for magnetic confinement in Cadarache, France, has a fusion chamber that weighs 5,639 tons (5,116 metric tons). Second, in 60 years of continuous, multipath research and development, no fusion reactor has ever produced usable energy. The application of fusion to space travel awaits the establishment of its basic power source.

THE BUSSARD FUSION RAMJET

Robert W. Bussard (1928–2007) was a fellow of the International Academy of Astronautics with a Ph.D. from Princeton University. In 1955, he joined the Nuclear Propulsion Division's Project Rover, designing and testing nuclear rocket engines. In 1960, anticipating a breakthrough in fusion power research, he designed the Bussard fusion ramjet for interstellar space travel. His engine is unique among all rocket concepts, in that the spacecraft using it for propulsion carries no fuel.

Interstellar space, while principally a hard vacuum, actually contains about 10^{-23} pounds per cubic foot (10^{-21} kg/m³) of matter, and it is primarily ionized and nonionized hydrogen with a small amount of helium-3. A large ramjet scoop, plowing through space at high speed, would have to sweep 4×10^{19} cubic feet (10^{18} m³) of space to collect 0.035 ounces (one gram) of hydrogen. In the Bussard engine, this collected hydrogen is then compressed into progressively smaller volumes defined by static magnetic fields, as driven by the high forward speed of the spacecraft. In the final compression, the hydrogen fuses, and the resulting energy is directed out the back of the ship, pushing it ever forward by the resulting thrust. For every gram of hydrogen collected, the fusion produces 1.8×10^8 watt-hours (6.4×10^{11} j) of energy.

The proposed scoop in front of the spacecraft is a virtual object, consisting of an invisible magnetic field, produced by machinery on board and shaped as a large funnel. To collect enough fuel to provide adequate acceleration, this magnetic scoop must be several thousand miles in diameter. Only some of the floating hydrogen atoms are ionized and can be influenced by the magnetic field to be driven into the engine, so a sweeping ultraviolet laser beam will pre-ionize the atoms and condition them to be guided in.

There are some issues to be worked out in the Bussard fusion ramjet. If the challenges of making a magnetic field thousands of miles wide and causing a continuous fusion reaction are set aside, there is the special problem of proton-proton fusion using hydrogen as fuel. It is true that the stars create enormous energy by fusing together simple hydrogen nuclei, but the reaction is actually very improbable. On average, it takes a billion years of trying before a hydrogen nucleus manages to fuse with another hydrogen nucleus, and it requires a mass of material the size of a star to create the necessary pressure and temperature to make the reaction possible. In the case of the Sun, which is an average star, this is 4.4×10^{30} pounds (2×10^{30} kg) of material. Even in a star, the fusion rate is close to zero. Seeing this as a sticking point, Bussard solved the problem by prescribing a *CNO* cycle for ramjet fusion, in which carbon, nitrogen, and oxygen are used as non-consumable catalysts in a recurring series of fusion interactions. The end product of the CNO cycle is an energetic helium-4 ion, which can be magnetically directed out the

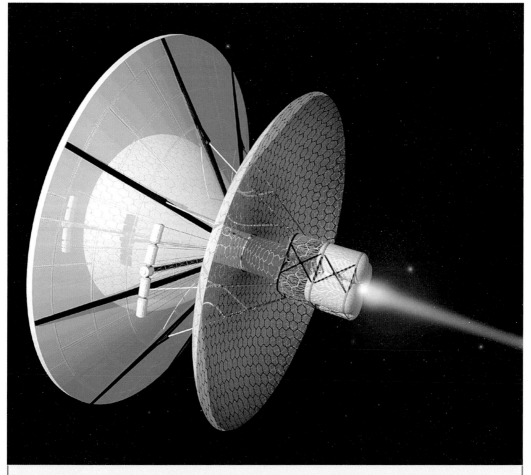

An artist's concept of what a Bussard ramjet engine may look like—an onboard laser is used to heat hydrogen plasma, and the laser or electron beam triggers fusion pulses, creating thrust. *(NASA Marshall Space Flight Center)*

exhaust nozzle, plus positrons and some energy lost by neutrino expulsion. In this process, the probability of fusion is increased by a factor of 10^{16}, but it usually occurs in stars that are 1.3 times the size of the Sun. A cyclic CNO fusion has never been accomplished in an Earth-bound laboratory.

There are also the practical problems of crossing the great divides of space separating stars. Using a reaction engine, such as the Bussard ramjet, it could take 100 years to reach a cruising speed of 77 percent of the speed of light. If it takes 100 years to accelerate to speed, then it will also take 100 years to come to a complete stop at the destination. To stop, the spacecraft must be turned completely around, with the engine operating against the trajectory of movement. It will be difficult to collect hydrogen in space with the funnel turned the wrong way.

PROJECT DAEDALUS:
AN UNMANNED FUSION ROCKET

Starting in 1973, the British Interplanetary Society of London spent five years designing an unmanned interstellar probe under the title Project Daedalus. The planetary system to be explored in this expedition is around Bernard's Star. This distance to be traveled, 5.9 light-years or 71 trillion miles, is small in interstellar terms. The trip will take 50 years, and the probe will streak past the target going 12 percent of the speed of light, disappearing forever into interstellar space. Observations will be conducted using two 16-foot (5-m) optical telescopes and two 66-foot (20-m) radio telescopes beginning at 25 years from passage, with data relayed back to Earth using a 131-foot (40-m) dish antenna, always pointed toward Earth, that doubles as the second stage engine bell. Once it reaches the target distance, it will take 5.9 years for the digital images to be received.

This proposed robotic probe is not small, with an overall length of 623 feet (190 m). It carries 55,000 tons (50,000 metric tons) of fuel and a payload that weighs 496 tons (450 metric tons). It is a two-stage rocket, to be launched from Earth orbit. The first stage rocket fires for 2.05 years with a continuous thrust of 1.7 million pounds (7.54 MN), accelerating the vehicle to 0.7 times the speed of light. The first stage is then separated and spends the rest of the mission as a radio relay back to Earth. The second stage fires, giving a 149,000-pound (0.663 MN) thrust for 1.76 years, after which the probe coasts at 0.12 times the speed of light for 46 years. To protect the spacecraft from hitting an occasional grain of sand at this speed, the front is protected by a disc of beryllium 0.27 inches (7 mm) thick, weighing 55 tons (50 metric tons).

Having a rocket engine that will burn for more than two years is an interesting challenge. A nuclear engine of some sort was the only choice, and for this probe it is a fusion rocket, fusing small, frozen pellets of a deuterium-helium-3 mix. Each pellet is to be coaxed into a fusion reaction, one at a time, in a chamber built to withstand the

THE NUCLEAR PHOTONIC ROCKET

The theoretical use of nuclear power for spacecraft propulsion opens many possibilities. One is the photonic rocket, which uses the thrust produced by radiation escaping from the ship in one direction, pushing it along. This is a rare example of a rocket that leaves no exhaust behind it. A nuclear fis-

shock of a small thermonuclear explosion. The pellet will be heated and compressed into fusion conditions using an aggressive electron beam, which will be powered by a set of induction coils in the rocket nozzle, taking advantage of the high-speed ions flying out the back of the ship. This rocket exhaust is a moving plasma, which is a superconducting material, and a conductor and magnetic field in relative motion generate electricity. To obtain the specified thrust, it will be necessary to inject 250 frozen pellets per second into the fusion chamber. Exhaust velocity from the first stage is 6,590 miles per second (10,600 km/s) and 5,720 miles per second (9,210 km/s) from the second stage.

Plans for interstellar exploration and fusion rockets in general tend to take inertial confinement fusion as a given technology, but repeated attempts to achieve such a reaction using the largest laser arrays on Earth have yet to report success. An even larger challenge may be the task of obtaining 28,000 tons (25,000 metric tons) of helium-3, which is an extraordinarily rare item. This problem has been addressed by the British Interplanetary Society. The strategy is to first park the Daedalus spacecraft in an orbit around the planet Jupiter. Several large hot air balloons will then be launched, each having a robotic helium-3 extraction factory suspended beneath, taking the fuel from solution in the Jovian atmosphere. These machines will have to operate continuously for 20 years to amass enough fuel for the Bernard's Star mission.

Daedalus is a grand exercise in engineering, intractable problem solving, and mission planning, but the greatest problem of all will be finding money to pay for the project. Other unlikely but well thought out nuclear engine designs include the redshift rocket and the fission sail, both of which were formulated for science fiction novels. The redshift is a variation of the antimatter rocket envisioned by novelist Karl Schroeder. The fission sail, written into a novel by Robert L. Forward, uses fission fragments to fill a large sail in the vacuum of space. Speculative stories may not seem a fertile ground for nuclear power concepts, but we must remember that the atomic bomb was first proposed in the novel *The World Set Free,* by H. G. Wells in 1914. That was 30 years before the first atomic bomb test in Alamogordo, New Mexico.

sion reactor could be operated at a sufficiently high temperature that it literally glows in the dark, and the light could be used as the propellant.

A less direct method of photonic propulsion, but one giving desirable control over the direction and collimation of the photon beam, is to use a fission reactor to provide power to a laser used as the photon source. Although there would be a loss of efficiency in using a laser over an incandescent reactor, the problems of unidirectional control are solved. A hot reactor would tend to radiate in all directions, and it would have to be located at the ray-intersection point of a large, heavy parabolic mirror to direct the photon beam in a constructive direction.

The power required from the reactor, assuming that an efficient laser could be designed, is 1.3 billion watts per pound (300 MW/N) of thrust. Current fission reactor designs can generate up to one kilowatt per pound (2.2 kW/kg) of reactor system. About 80 percent of the mass of the vehicle powered by this rocket would have to be fissile reactor fuel, so if the spacecraft weighs 660,000 pounds (300,000 kg) then 530,000 pounds (240,000 kg) of it is pure plutonium-239. When the fuel load is completely burned out, the ship will weigh 530 pounds (240 kg) less than it did before the mission, without having shed any exhaust. The lost weight is from a pure mass-to-energy conversion.

The nuclear photon rocket ship will accelerate at a very leisurely 10^{-5} gravities (0.1 mm/s^2). Running at full power, it will gain 9,800 feet per second (3,000 m/sec) per year. This is sufficient acceleration to provide interplanetary travel from low Earth orbit. The rocket will have to accelerate for a year to escape Earth's gravity. In this hypothetical case, it would take 80 years of steady photonic thrust to reach a reasonable interplanetary velocity of 150 miles per second (240 km/s). At this rate, a nuclear photon rocket mission is probably best conducted robotically.

ANTIMATTER ROCKETS AND ANTIMATTER CATALYZED NUCLEAR PULSE PROPULSION

The ultimate nuclear-powered rocket, the design having the highest energy density and the most specific impulse, may be the antimatter rocket. The energy density of antimatter as fuel is more than 10 billion times greater than the best chemical rocket fuel and 100 times better than the most efficient theoretical fusion rocket.

The antimatter rocket carries two types of fuel: matter and antimatter. If ionized hydrogen (protons) is used as the fuel, then an equal amount of ionized anti-hydrogen (antiprotons) is carried. Allow a small portion of protons to come into contact with an identical portion of antiprotons, and the two annihilate each other with extreme violence. Unlike fission, which converts only 0.1 percent of the fuel mass to energy, an antimatter reaction takes it all. If this action can be directed out the back of the vehicle, then forward motion can be established using a minimum fuel load.

All subatomic particles, electrons, protons, and neutrons, have their anti-equivalent, but the proton, the nucleus of hydrogen, is probably the best for antimatter annihilation in a rocket motor. The energy conversion of proton-antiproton annihilation is complete but far from perfect for propulsion purposes. About half of the energy is taken away in neutrinos and is completely unrecoverable. About half of the recoverable energy is in charged mesons, and half is in uncharged mesons. The charged mesons can be used to power a space vehicle, because their direction of travel can be controlled using magnetic fields. The uncharged mesons quickly decay into gamma rays shooting off in random directions. There is no way to control a gamma ray or use a mirror to make it do anything but scatter, so its energy causes zero headway for the rocket, and shielding must be used to keep gamma rays from sterilizing the spacecraft.

Another possibility, the electron-positron interaction, simply wipes out the two particles and converts them into gamma rays, and there is no direct use for this energy. However, the intense gamma rays emitted by electron-positron annihilation could be used to heat a solid engine core. Hydrogen gas emptied into the core would expand rapidly and leave the ship through a nozzle. This method of energy exploitation requires that liquid fuel be carried, and this negates some of the advantage to using antimatter.

A rocket using proton-antiproton direct thrust would require only 10 grams of antimatter to reach Mars in one month. The antiprotons, and separately stored protons, would be formed into frozen pellets. The antimatter would have to be stored in suspension, not touching any matter until the moment of propulsive destruction, using magnetic levitation. No special ignition machinery is necessary. Throw two frozen pellets at each other, one protons and one antiprotons, and they will self-destruct. There is no question about this process.

While possible in theory, the major stumbling block for this rocket design is identification of an antiproton source. Antiprotons are made, one at a time, using high-powered particle accelerators simulating cosmic rays crashing into the upper atmosphere. As much power has to go into

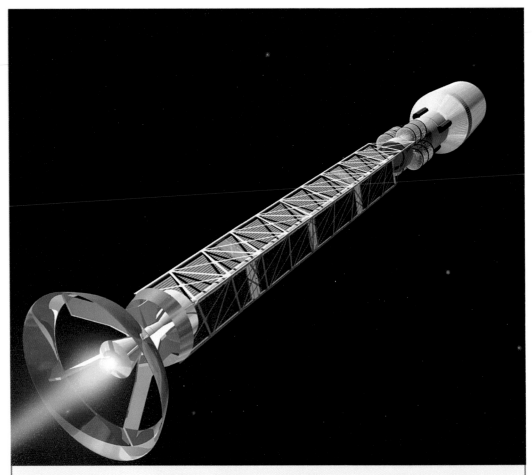

An artist's concept of an antimatter propulsion system *(NASA Marshall Space Flight Center)*

making an antiproton as can be derived from its annihilation, assuming that the process is 100 percent efficient. The process of antimatter production is, however, extremely inefficient, with probably one unit of energy usefully converted for every 1 million units used. The manufacture of antimatter rocket fuel is therefore impractical, given our current means.

There is another way to use antimatter to advantage for nuclear rockets, called antimatter catalyzed nuclear pulse propulsion. Nuclear pulse propulsion was first proposed in the late 1940s as a very efficient means of powering a spaceship. The concept was simple, and it was possible to implement using only the technology that was available at the time. A spacecraft is made spherical, with one opening, which is a rocket nozzle.

It is propelled forward by setting off nuclear bombs, one at a time, at the center of the sphere. The entire bomb is reduced instantly upon detonation into energetic plasma at extremely high temperature. It expands in the sphere and escapes through the nozzle, producing a pulse of thrust. Over and over new bombs are pulled to the center of the sphere and exploded, with each event giving a push and an increment of velocity.

The effect is crude but sure to work. The only problem is the size of the explosion. A nuclear weapon makes a tremendous blast, the strength of which is expressed as equivalent to tons of TNT. It is difficult to scale this explosion down to a reasonable level, in which the entire spacecraft is not torn asunder or even reduced to vapor. A nuclear bomb is a hypercritical mass of pure, fissile material. In theory, one can be built with a yield as small as 10 tons (42 GJ). Below that level an efficiently burning critical mass is impossible to assemble. What is needed for a nuclear pulse rocket is a small but efficient explosion, using a fissile fuel such as uranium-235 or plutonium-239. An explosion no larger than one ton of TNT would be manageable, allowing a strongly built spacecraft to remain in one piece as hundreds of explosions are used to accelerate it to cruising speed.

A way to initiate multiple fissions in a fissile material, simulating a nuclear bomb explosion on a smaller scale, is to introduce a sudden spray of neutrons from an external source. If done right, this will cause trillions of fissions at once and excite the subcritical mass into destruction, as if it were a bomb. Many ways of accomplishing this effect have been tried, including a hydrogen fusion neutron source. A most effective method of spraying a bomb core with neutrons is to use antiprotons. An antiproton has a negative charge, just like an electron. If introduced into a matrix of uranium atoms, an antiproton will be captured by the positively charged nucleus of the first one it encounters. Spiraling down into the heavy nucleus in a rapidly decaying orbit, the antiproton will eventually touch one of the 92 protons, and the two particles will annihilate one another. The force of the explosion will rip the nucleus to pieces, sending the 143 neutrons flying and causing nuclear fissions when they come in contact with adjacent nuclei. The result of a single proton-antiproton interaction is a very efficient spray of neutrons. Multiple, simultaneous proton-antiproton destructions cause a simulated nuclear bomb explosion, scaled down to usable size.

In 1990, an interplanetary spacecraft named *ICAN-II* using antimatter catalyzed fission was designed at Penn State University. It is capable of sending a manned expedition to Mars, with a transit time of only 30 days. The trip would require only 5×10^{-9} ounces (140 ng) of antiprotons, in addition to the fissile material. This small amount of antiproton material, combined with the known properties of fission, brings the concept of antimatter propulsion closer to reality.

THE GAS CORE REACTOR ROCKET

When considering a nuclear fission reactor for power production or propulsion, it is often assumed that the core of the reactor, the cylindrical assembly containing uranium at the center, is in danger of melting if the reactor temperature is allowed to rise above the melting point of the material. This factor limits the amount of power that can be derived from a reactor. It must be run at a power low enough and a cooling rate that is high enough to prevent melting. However, there is another way of looking at the problem of producing power with fission. Suppose a reactor could be run at an extremely high temperature without the worry of melting. If the core were gaseous and not solid, then melting would not be a problem.

It is possible to achieve a critical mass and run at extremely high power, using uranium hexafluoride gas as the fuel. This gas is contained in a cylinder, with a neutron moderator made of beryllium oxide surrounding it. The fission reaction is controlled by varying the amount of beryllium oxide that faces the core. The uranium hexafluoride can be made supercritical, raised to a power level, and then leveled out at exact criticality to start heating the reaction fuel. The core of such a reactor is expected to run at a temperature of about 45,000°F (25,000°C). At this temperature, the reactor is operating beyond white-hot. The main component from the thermal radiation is ultraviolet. Expose a propellant, such as liquid hydrogen, to this radiation, and it will expand explosively. The propellant is separated from the reactor core by a heat exchanger made of pure quartz. Direct the force of the hot, expanding gas through an exhaust nozzle, and the spacecraft moves quickly by reaction in the opposite direction. The gas core reactor gives a specific impulse in the range of 3,000 to 5,000 seconds (30 to 50 kN·s/kg) and enough thrust for manned interplanetary travel. Travel times to outer planets can be collapsed from decades with chemical rockets to weeks.

As is the case with many nuclear propulsion plans, in theory it makes good sense, but the devil is in the details. There is no problem with a melting core, but there is the problem of the container in which this reaction can take place. Uranium hexafluoride is an extremely reactive chemical and will corrode almost everything. Heat it to ultraviolet temperature and no material is known that will contain it. Magnetic bottle technology has been suggested, because the reactor core is entirely ionized and could be treated as a plasma, but the outward pressure exerted by this plasma will be extreme, on the order of 1,000 atmospheres (100MPa). A magnetic field to contain this pressure will require a massive superconducting magnet. The weight of the magnet may negate the advantages of the gas core reactor.

THE NUCLEAR ELECTRIC ROCKET

Rocket propulsion generally involves accelerating a propellant out of the vehicle in one direction. The spacecraft will then move in the direction opposite that of the exhaust. This system of propulsion works well in the vacuum of deep space. As long as material in the rocket is accelerated away from it, it will gain speed until relativistic effects begin to affect the acceleration. The speed of light cannot be reached.

A problem with traveling the extreme distances between star systems is that a great deal of heavy fuel cannot be carried for such long trips. Early in the speculative designs of interstellar spacecraft in the 1950s, the need was seen for extremely efficient rockets using a minimum amount of fuel. There are two ways to maximize the amount of forward force, or thrust, produced by the rocket. Either push a lot of mass out the exhaust nozzle or push a little bit of mass very quickly. To achieve the second goal, the ion rocket was invented.

The ion rocket is purely electrical. A fuel, such as ionized xenon gas, is injected into the rocket engine through a hole in a positively charged anode. Drawn by the powerful electrostatic pull of the negative charge on a cathode, the xenon is accelerated to escape velocity and leaves the vehicle through the rocket nozzle. An electric current, running through the xenon plasma from anode to cathode, is supplied by an electric generator. All of the energy used to eject the fuel and thus accelerate the spacecraft is electrical.

This method of space propulsion has been developed and tested in flight, beginning with NASA's Space Electric Rocket Test, SERT-I, on July 20, 1964. On September 27, 2007, the interplanetary probe *Dawn* was launched by NASA, and its mission is to explore the asteroid Vesta and the dwarf planet Ceres. It uses three NSTAR xenon thrusters, each developing a specific impulse of 3,300 seconds (32 kN·s/kg) using 2.3 kilowatts of electrical power. The NSTAR engines have been tested for continuous operation at full power for 3.5 years.

The ion-drive rockets have been developed, tested, and used in multiple space missions, and this may be the correct technology for a deep space or interstellar mission. However, for such a trip, solar panels cannot be relied on for power. There is simply no usable sunlight in the space between stars. Nuclear power reactors will be relied on to provide the electricity. Several types have been investigated, including a specially built pebble bed design. The fuel planned for such a long-term mission, probably more than 100 years, will be highly enriched uranium pellets encased in graphite balls, with a touch of boron added. The reactor will heat nitrogen gas, which will then operate a gas turbine, which spins an electrical generator. A great deal of radiator surface will be needed on the vehicle to recool the nitrogen for further use and expel the excess heat from the turbine cycle, sending the heat away into the blackness of space.

The purpose of the boron added to the fuel is to suppress the fission action. As the years go by, the fuel will burn up, and the reactor core will lose its ability to maintain a self-sustaining reaction. However, the boron will also become ineffective at preventing the process, as it activates by neutron capture into an isotope that is not likely to affect fission. The reactor is given new life, and the mission continues on without a need to refuel the reactor.

THE NUCLEAR SALT-WATER ROCKET

The proposed nuclear rockets discussed to this point have been suitable for long-distance traveling. For missions to the outer planets or the stars, a steady thrust applied for many years is appropriate, and efficiency and minimum fuel mass are most important. However, the thrust values from these technologies are very low and do not supply the type of heavy lift that will get a rocket off the ground or blast it to the Moon. For this reason, the hard work of space travel has always been done by solid or liquid fuel rockets. They are extremely inefficient, and the fuel is gone after a few minutes of running, but they get the job done using chemical burning.

A fission engine that combines the brute force thrust potential of a chemical rocket with the ideal energy efficiency of a nuclear rocket has been designed. It achieves tremendous thrust with a continuous nuclear explosion. Uranium-235 or plutonium-239, both of which have been used as the active ingredient in atomic bombs, is configured in a hypercritical, prompt-fission situation. The fission reactions run out of control. The resulting explosion is initially contained in a stout chamber, with the instantly expanding explosive gases and plasma exiting through an exhaust nozzle. The explosion continues outside the spacecraft, adding to the net thrust resulting from the violent nuclear reactions, and the ingredients for further explosions are continuously fed into the engine. It is called the nuclear salt-water reactor (NSWR).

The fuel for this engine is salt water, but the dissolved salt is a chemical compound of either uranium-235 or plutonium-239. The water is a medium for holding the fissile fuel in tanks, pumping it easily through pipes, and injecting it into the reaction chamber. It also becomes part of the mass that is ejected from the exhaust nozzle at high speed, causing forward thrust. Energy release from the fission is at least a million times greater per unit mass of fuel than can be achieved with chemical burning. The fuel must be kept in special tanks to prevent it from going critical. Tanks can be long and thin, with boron walls to absorb fission-causing neutrons. Once it is injected into the engine chamber, the salt water is given a geometry very favorable to fission with neutron reflecting walls, and the reaction becomes explosive.

Such an engine could develop 3 million pounds (13 MN) of thrust and an exhaust velocity of 41 miles per second (66 km/sec). The best chemical rockets can achieve an exhaust velocity of only about 2.7 miles per second (4.5 km/s). An optimized nuclear salt-water rocket could develop an exhaust velocity as high as 2,900 miles per second (4,700 km/s) and accelerate a 330 ton (300 metric tons) payload to 3.6 percent of the speed of light using 3,000 tons (2,700 metric tons) of saltwater fuel.

The nuclear salt-water rocket could easily lift off from a planet's surface and proceed directly to another planet or moon. However, the exhaust is highly radioactive, containing fission products and still fissioning fuel in an expanding cloud of superheated steam. The safety and environmental pollution factors are severe. The danger factor may limit the use of an otherwise interesting application of fission.

THE RADIOISOTOPE ROCKET

Early in the 20th century, the radioactive element radium was discovered, and shortly thereafter new theories of relativity explained the release of radiation in terms of a direct conversion from mass to energy. It was a rare metal, occurring in trace quantities in uranium ore, but a few milligrams could produce measureable heat and light up the corner of a room. This energy release happened without combustion and the presence of air. Speculative applications of this potential windfall were pondered and published.

An obvious use for radium was to power submarines. Submarines suffered from the fact that there was no air available underwater, and therefore they could run only on battery-powered electric motors. The top speed and range of a submarine below the surface was thus severely limited. A few pounds of radium, on the other hand, placed at the bottom of a tank of water would cause it to boil and make usable steam. The steam could drive a turbine connected to the propellers. A submarine would never want for fuel, because radium-226, the common isotope, has a half-life of 1,600 years, and the engine could run in a completely sealed environment. No air was needed. There was only one overriding problem. In the entire world, there was less than an ounce of radium-226 separated from uranium ore. It was the world's first nuclear propulsion concept. It was a logical idea, and, as seems the case for many such ideas, the only reason it would not work was the utter lack of the active ingredient.

After World War II, isotope production reactors were manufacturing radioactive isotopes on a large industrial scale, and the need for the rare radium-226 was lost. Isotopes for multiple applications were available in usable quantities at realistic costs. With the beginning of space exploration programs in the United States and the Soviet Union in 1957, the old concept of making steam with radium was dusted off. With the heat of a fixed isotope heat source, it would be possible to flash-boil a liquid propellant, such as liquid hydrogen, and direct the hot gas through an exhaust nozzle. It would make an energetic, simple rocket engine that uses only one fuel component. Temperatures from 2,700° to 3,600°F (1,500° to 2,000°C) are possible, producing respectable specific impulse values from 700 to 800 seconds (7 to 8 kN·s/kg). This is about double the performance of the best chemical rocket engines ever made.

The available power level is fairly low. While a fission reactor can produce 1 billion watts of power on demand, a fixed isotope mass of reasonable size gives about 5,000 watts. This means thrust levels of 0.29 to 0.34 pounds (1.3 to 1.5 N), making it useful for long-term, unmanned missions or for use as attitude thrusters. The most often used heat-producing isotope is plutonium-238, with a half-life of 87.8 years. It has been applied to deep space programs for several decades, producing electrical power out of the range of solar panels and keeping the electronics warm in all spacecraft sent to Jupiter and beyond. Greater power can be derived from very short-lived isotopes such as polonium-210, with a half-life of 138 days.

In 1961, the now defunct aerospace company TRW began a development program named Poodle, working on a radioisotope rocket. By April 1965, a completed engine was run under test for 65 straight hours at about 2,700°F (1,500°C). As predicted, its specific impulse was about 700 seconds (7 kN·s/kg). Today, the radioisotope units used in spacecraft are known as poodle thrusters.

Someday, when mankind gets the urge to explore the greater solar system and beyond to the stars, these ideas will be here for further research and development. Only one thing is certain about the propulsion system that will be used, first for unmanned and then for manned spacecraft: The system will gain its power from a nuclear reaction. We have actually been closer to this realization than many realize. In the next chapter, the story of the quest for practical, working nuclear-powered rockets and jet engines is revealed.

6 The Future as Seen in Past Projects

The nuclear rockets surveyed in the previous chapter may seem fanciful long shots or something that may happen in the distant future. This may be true, but rocket and jet propulsion were pursued in dead earnest in the 1950s and the 1960s, with the purpose of sending men to Mars, Jupiter, and Saturn, and establishing a shuttle service to a permanent Moon base. Other priorities eventually diverted our attention and our money to other pursuits, but for a while the U.S. government had big plans for space exploration, and it would obviously involve nuclear power. Reactors were built and tested to go into orbit, supplying power for satellites with high-energy needs, and plans were to put a nuclear power plant on the Moon, to provide all the power needed for human habitation. High-thrust nuclear rocket engines were designed, tested, and perfected in desert locations far from civilization.

The fact that no application came into being for these devices does not mean that effort was wasted. These propulsion systems, developed out of the public eye, may coincide with a renewed interest in interplanetary and interstellar transportation as mankind strives to find other life in the universe.

Speculation and faraway plans for nuclear rockets and jet engines began within days after the first nuclear reactor attained criticality in Chicago on December 2, 1942. The possibilities for the use of this new energy source, with its compact efficiency and no need for air, were seemingly endless, but first World War II had to be won, so all attention was turned to designing and building nuclear weapons. With the Japanese surrender on August 15, 1945, other applications for nuclear fission could be considered. Among the first ideas to be addressed after the war with a federally funded program was the nuclear-powered strategic bomber. Together with nuclear-powered submarines, nuclear bombers would ensure that the United States could neutralize potential enemies with an umbrella of nuclear weapons held constantly in the air and ready to strike any point on Earth with a few hours' notice. Using nuclear engines, bombers would be able to stay aloft for months, circling, while nuclear submarines stayed below the oceans, always at the ready. Multiple preliminary studies indicated that nuclear jet engines were practical, and a joint project for the air force and the Atomic Energy Commission (AEC), the Aircraft Nuclear Propulsion (ANP) program, began in 1951.

Laboratories and test facilities were built all over the country, from Idaho to Georgia, and billions of dollars were expended, working on everything from robots to remotely service the engines to in-flight meals planned for bomber crews. A bomber equipped with a working nuclear power reactor was flown between Texas and California 43 times. This expensive project was cancelled in 1961, before an operational nuclear bomber was put into service. Intercontinental ballistic missiles, which had been developed in a parallel program, rendered the more expensive manned bombers obsolete, and the program was cut from the federal budget.

It was just as well that the project was killed, because the public safety issues of flying an airplane with naked nuclear reactors on the wings was beginning to cause concern. On the ground or underwater, reactors are easily shielded against radiation emission using heavy materials, such as concrete, steel, and lead. An airplane, on the other hand, had to be made light enough to fly. There was fear that the bomber crew would perish from the radiation, and a crash could sterilize many square miles of ground. Still, the overpowering need for the extreme, concentrated power of nuclear fission in propulsion systems meant that the concept was pursued in other programs. This chapter covers the Orion Project, a rational design for long-distance space travel, the NERVA advanced nuclear rocket engines, and a project conducted in the deepest secrecy, a cruise missile with a nuclear ramjet. Sidebars discuss Freeman Dyson, one of many gifted scientists who worked on these projects, and Project Longshot, a planned trip to Alpha Centauri.

THE ORION PROJECT: PROPULSION BY NUCLEAR EXPLOSIVES

In 1947, Stanislaw Ulam (1909–84), a multitalented Polish mathematician, had finished work on the Manhattan Project at Los Alamos, New Mexico, and was ready for new challenges. Using the successfully designed atomic bomb as a starting point, he conceived a fission-powered interstellar spacecraft. The vehicle would simply be pushed along, in sharp jolts, by exploding nuclear weapons, thrown out the back of the ship, one at a time. In the vacuum of space, a small bomb would be shot out the back using a compressed air cannon. When it reached a certain distance away, it explodes. The shock wave developed in the atomized debris hits the back of the vehicle and jolts it forward, giving it an incremental acceleration. Do this enough times and the speed of the spacecraft can eventually get to within 80 percent of the speed of light. Ulam, whose life's accomplishments would include the hydrogen bomb, considered this idea to be his best. It was unique. It was a design for an interplanetary or even interstellar spacecraft that could be built with existing materials and techniques. There was no need to develop any nonexistent technology, find any extremely rare fuel, or use any superconducting magnets that are 100 miles wide. What is more, it could take off from Earth, blow through the atmosphere, travel 62 million miles (100 million km) to Mars in a few weeks, off-load a base of operations with provisions, and return to Earth using exactly one rocket stage. There would be no need to jettison large chunks of the vehicle as it blasts off the ground and flies to Mars. On paper, it was a very attractive proposition.

Freeman Dyson (1923–) a British theoretician of similar talents, had worked on his own nuclear spacecraft design using atomic bombs for fuel in a funded project called Helios. As was the case with Ulam's project, the work was classified secret, due to the fact that it involved nuclear weapons. His proposed spacecraft was a sphere, 130 feet (40 m) in diameter. To produce a single jolt forward, a small, 0.1-kiloton (4×10^{11} j) atomic bomb is set off in the center of the sphere, and the vapor from the fully consumed bomb exits through an exhaust nozzle. To increase the thrust, water is first injected into the chamber with the bomb, and its resulting superheated steam gives an enhanced push. Dyson calculated the specific impulse to be about 1,150 seconds (1.1 kN·s/kg), which was more than twice the performance of the best chemical rocket.

Dyson eventually bumped into Ulam's nuclear propulsion design, and he had to admit that it was much better. Helios was shelved, but in 1958 Ted

Taylor (1925–2004), a most accomplished fission bomb designer, quit work at Los Alamos and signed on at General Atomics in San Diego, California. General Atomics was a major defense contractor, working primarily in nuclear physics. Taylor started a secret program for the air force called Project Orion. The goal was to fully develop a large single-stage, nuclear-powered spacecraft, based on Ulam's original design. Dyson took a year's leave from his prestigious post at Princeton's Institute for Advanced Study to work on the project.

The concept of riding in a vehicle being pushed along by atomic bombs in close proximity is not as questionable as it may sound. The midrange *Orion* spacecraft was to be about 130 feet (40 m) long, with the passengers in the front. The bombs were detonated 200 feet (60 m) in back, and the shock wave was caught with a thick, round steel pusher plate. The passengers were shielded from any radiation burst from the bomb by the distance from the blast and the mechanical structure of the vehicle. The jolt,

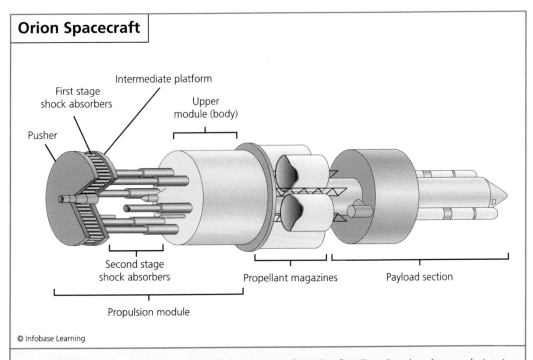

Orion Spacecraft

A schematic diagram showing how the *Orion* spacecraft works. Small nuclear bombs are shot out, one at a time, through the center of the pusher plate. They explode in space at a short distance away, and the expanding sphere of gas created by the explosion hits the pusher plate, bumping the ship forward.

which would be severe enough to smash dishes in the dining area, was softened using two stages of heavy shock absorber arrays, taking up about one-quarter of the length of the spacecraft. These shock absorbers would not eliminate the forceful jolt of the bomb or waste power doing so, they would only spread the shock out over time. A 4,000-ton (3,600 metric ton) vehicle would have a speed increment of 20 miles per hour (10 m/s) from each detonated atomic bomb, spaced at half-second intervals. This acceleration over the 3-millisecond duration of each explosion would be dangerously bone-jarring and would tend to disassemble the spacecraft, but if spread over half a second by the shock absorber, each jolt in the forward cabin is softened to an acceleration of less then two gravities (19.6 m/s^2).

In 1959 it was time for testing these bold designs. A scale model of the *Orion* spacecraft, three feet (one meter) in diameter, was mounted on a launch gantry in a secluded spot on Point Loma, a peninsula in San Diego harbor. The nuclear-powered Orion was meant to blast off from a standing start on the ground, and this was considered the most difficult phase of a flight. The concept of being lifted into the air by a nuclear explosive underneath was disquieting, and first the basic concept needed to be proven using conventional explosives. Small bombs made of C-4 military explosive were stacked vertically in the model, to be pushed through a hole in the middle of the steel pusher plate and detonated individually, in sequence. The experimental series was code-named Putt-Putt.

The experimental results were encouraging, with the bullet-shaped model lifting off, as if it were powered by a conventional rocket, and accelerating vertically. The test vehicles were recovered by parachute. With the first question out of the way, the next concerned the pusher plate. The radiation from each nuclear explosion would be primarily ultraviolet light, but the plasma ball from it would be heated to 120,000°F (67,000°C) when it hit the steel disc. There was a question as to whether the pusher plate could withstand this assault, over and over.

Research uncovered a previously conducted experiment. On August 27, 1957, in Operation Plumbbob at the Nevada nuclear weapons test site, a small-yield bomb named Pascal B had been detonated at the bottom of a vertical shaft. Atop the shaft was a 1,980-pound (900-kg) steel plate. High-speed motion pictures were made of the shot, and one frame of the movie showed the disc moving rapidly upward. Calculations indicated that it had accelerated to six times the Earth escape velocity, and if it survived it should have been halfway to Mars by then. Further consideration of its speed indicated that it had most likely burned up from air friction,

FREEMAN DYSON (1923–):
THE MAN WHO SEES THE FUTURE

Among his many other technical accomplishments, Freeman John Dyson was instrumental in the development of the *Orion* spacecraft, which was to be pushed along by atomic bomb explosions. Dyson was born on December 15, 1923, in Crowthorne, Berkshire, England, son of the English composer George Dyson. The fact that he shared the same last name with a famous astronomer, Frank Watson Dyson (1868–1939), sparked his interest in science. The two were not related.

Dyson found work as an analyst for the RAF Bomber Command at RAF Wyton military airfield near Cambridgeshire during World War II. In his work at Wyton, he was credited with creating the field of operational research, in which the use of technology by an organization is optimized by mathematical means. Using his experience during the war, he was able to earn a B.A. in mathematics at Cambridge University in 1945 and became a fellow of Trinity College, Cambridge, in 1946. In 1947, he moved to the United States on a fellowship at Cornell University. Concentrating on the most challenging aspects of nuclear physics, in 1949 he demonstrated proficiency in quantum electrodynamics theory. In 1951, he became possibly the only physics professors at Cornell without a Ph.D. and without formal training in physics. His intense work in mathematics included such topics as topology, functional analysis, number theory, and random matrices.

In 1957, he went to work at General Atomics in San Diego on an exciting, secret program for the air force, the Orion Project. The next year he also split his time on another important project at General Atomics, leading the design team for the TRIGA reactor. The TRIGA was a unique nuclear reactor design, in that it was impossible to melt the core no matter how badly it was mistreated. In 1961, the Orion Project came to a close because of the new treaty that banned aboveground nuclear weapons testing, and Dyson moved on to equally exciting programs. In 1966, he proved that the quantum exclusion principle plays the main role in the stability of bulk matter. This work answered a fundamental question: If one block of wood is placed atop another block of wood, why do they not coalesce into one block of wood? Obviously they do not, but why? It had been assumed that the natural repulsion of electrons in the outer reaches of atoms in the wood repelled similar electrons in other pieces of wood, but this turned out not to be the case. As proven, a phenomenon called the classical

(continues)

(continued) _____

macroscopic force prevents one chunk of matter from naturally merging with another chunk of matter.

For his prodigious work in physics, Dyson was awarded the Lorentz Medal in 1966, the Max Planck Medal in 1969, and the Templeton Prize in 2000. Working at Duke University and for Princeton's Institute for Advanced Study, he formulated such mathematical inventions as the Dyson sphere and the Dyson operator. He may agree with many of the scientists who worked on the Orion Project that it was the best time of their lives. The careful designs that came from the engineers on the project convinced him that an *Orion* spacecraft would put people on Mars by 1965 and in orbit around Saturn by 1970, and he was deeply disappointed when the project was stopped and sealed shut. He still hopes that inexpensive space travel, not requiring the resources of an entire government, will come about later in this century. The problem with our lack of interplanetary travels, he believes, is risk aversion. If we cannot take a chance on failure, then we will never get there. Dyson could probably prove this mathematically if he wanted to.

being the reverse of an iron meteor entering the atmosphere at high speed. Nuclear tests specifically designed for *Orion* were conducted, and Ted Taylor designed a special bomb just for use in propelling the vehicle.

In testing, steel plates seemed to survive an atomic blast more or less intact, but the surfaces eroded away, as if sandblasted. This would limit the lifetime of a critical component of the *Orion,* but a solution was found by accident. In an early test, someone had left fingerprints on a pusher plate. Under the fingerprints, the steel was unharmed after a full-scale test. Nothing but the oil in a fingerprint had protected the metal from a nuclear bomb explosion at close range. Further tests were made, with pusher plates covered with oil. It worked perfectly, and a revised design of *Orion* included a central grease gun to spread lubricant over the surface before each explosion.

With these encouraging results, the project entered a design and planning stage for interplanetary exploration. Three *Orion*s were designed: a satellite version for orbital missions, a midrange to go to the moon, and a super, weighing 8 million tons (9 million metric tons) and 1,300 feet (400 m) in diameter. The super version would be capable of landing a permanent base on the Moon in one trip. With the good results came disadvantages. As it blasted off from the ground, the *Orion*'s bomb explosions would leave radioactive fallout in the atmosphere. Dyson calculated that a heavy liftoff would be the equivalent of a 10-megaton (40-pj) aboveground nuclear blast, and it could, in statistical terms, lead to one fatal cancer from the fallout. Assembling an *Orion* in orbit and then using the bomb detonations to escape would make little difference. The heavy dust from the bombs would slowly sift down through the atmosphere, and the end result would be the same, with a rise in the Earth's background radiation.

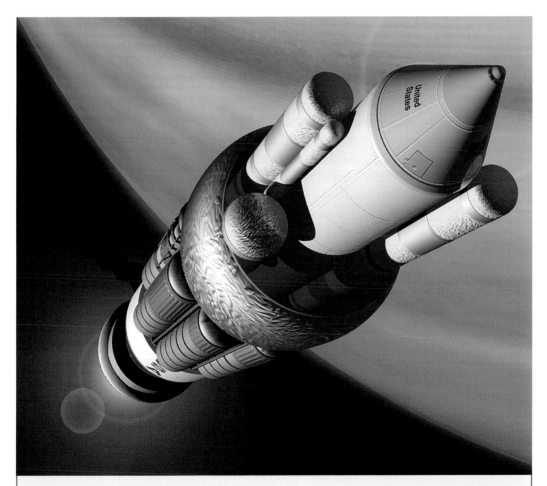

An illustration of the pulsed nuclear fission propulsion system concept investigated under Project Orion—the spacecraft is shown orbiting Jupiter, with human occupants. *(NASA Marshall Space Flight Center)*

The project came to an abrupt end in 1963, with adoption of the terms of the Partial Nuclear Test Ban Treaty with the Soviet Union and the United Kingdom. This treaty, intended to slow the nuclear arms race and stop the salting of the Earth with radioactive fallout, specifically banned aboveground nuclear tests, even if they were not intended as weapons. *Orion* could not be tested in development or flown upon implementation. The project office closed shop, and most documentation was classified secret.

Project Orion, although its premise seems unlikely, remains a well-considered, well-designed spaceflight option that is based on existing technology. It may be dusted off in the future and used to send mankind to the planets and, eventually, to the stars.

NERVA AND KIWI: NUCLEAR THERMAL ROCKETS FOR THE *APOLLO* BOOSTER

In 1952, several applications for a controlled nuclear power source were considered for further development. A nuclear-powered submarine seemed a logical choice and so did a nuclear rocket engine, in anticipation of outer space exploration. The Los Alamos Scientific Laboratory in New Mexico began researching the topic. Nuclear reactors in the early 1950s were large and heavy, the size and weight of entire buildings, and were not necessarily suitable for weight-conscious rocket designs. By 1955, the Lawrence Livermore National Laboratory was able to shrink the size and weight of a reactor to something usable, a cylinder the size of a garbage can, and Project Rover began.

Project Rover was an aggressive program to build practical rocket engines that could exploit the extremely favorable energy density in a compact nuclear fission reactor. Nuclear rockets were not just designed on paper in this project. They were designed, built, rolled out to the test stand in Jackass Flats, Nevada, started up, and run to full power, blowing nuclear exhaust into the air and the ground. Test buildings were even made that could have the air pumped out, testing nuclear rocket engines in a simulated environment of outer space. The tests and experiments were so successful and promising that by 1961 NASA's Marshall Space Flight Center in Alabama began planning missions using nuclear engines. The *Space Nuclear Propulsion Office (SNPO)* was opened jointly by NASA and the AEC to manage the rapidly developing nuclear age of spaceflight. Strict objectives and nuclear engine specifications were drawn up, and improvements of the nuclear rocket performance continued to impress observers at the Nevada test site. A *reactor in-flight test (RIFT)* was on the schedule for 1964.

Many engine designs were built and tested, with problems of reactor core design and materials discovered and corrected with each test. The nuclear rocket engine concept is not very complicated. Fuel, such as liquid hydrogen, is pumped into a small reactor running under exactly critical conditions at high power. The fuel flashes into superheated vapor and leaves the reactor through a narrow nozzle, causing a high-speed exhaust that thrusts forward the engine and the vehicle to which it is attached. The heat energy imparted to the hydrogen is at least twice the excitation available from combining it explosively with liquid oxygen, and the result is an unusually efficient rocket engine that does not need the extra weight of a liquid oxygen supply.

Cylindrical nuclear reactor cores were constructed using highly enriched fuel. A rocket reactor must be made of extremely tough ceramic material, as the temperature range over the core, which is only 4.3 feet (1.3 m) long, is extreme. At the liquid hydrogen fuel inlet, the core is cooled to -420°F (-251°C), and at the exhaust end the core is heated to 4,941°F (2,727°C). Hydrogen in the reactor core acts as an efficient neutron moderator, allowing the reactor to achieve and maintain criticality as long as there is hydrogen fuel running through it. The reactor shuts down automatically when the fuel stops flowing. The liquid hydrogen is pumped from a cryogenic tank into the top of the reactor core using a turbo pump. The nozzle, which is made of metal, must be kept from melting in the extremely hot exhaust by cold liquid hydrogen running through coils attached to the outside. The heated, expanding hydrogen gas is routed back to a turbine, which turns the fuel pump. A large nuclear rocket engine is typically only 22 feet (6.7 m) long, including the fuel pump and the large nozzle skirt, yet it can develop 75,000 pounds (334 kN) of thrust with a specific impulse of 875 seconds (8.58 kN·s/kg). This performance is twice what has ever been achieved by a rocket burning liquid hydrogen with liquid oxygen.

The nuclear rocket engine tests beginning in July 1959 were codenamed KIWI. The first engine on the stand was KIWI 1, and it ran at a modest power of 70 MW, testing the uranium oxide core for durability in the extremely corrosive liquid hydrogen. Further tests were KIWI A Prime and A3, using coatings on the fuel to improve the wear characteristics. The KIWI Series B tests used fuel pebbles coated with niobium

The KIWI-A Prime nuclear rocket engine, sitting vertically on the test stand *(NASA)*

carbide. The fuel was changed to uranium carbide for the last KIWI tests in 1964. Problems of fuel erosion and cracking were largely solved in the KIWI test series. In every test of a nuclear rocket, fuel particles tended to break off the core and fall to the ground downwind of the test. Much effort went into minimizing this effect.

Using the KIWI data and experiences, a series of larger engines named Phoebus were built and tested at extremely high power. In February 1967, Phoebus B ran for 30 minutes at 1.5 billion watts. The final Phoebus 2A test ran in June 1968, producing 4 billion watts of heat for more than 12 minutes. Phoebus 2A was the most powerful nuclear reactor ever built.

KIWI and Phoebus were never built to fly, but in 1964 testing began on NERVA NRX, an engine built to SNPO specifications for manned space-flight. Its stated purpose was to power an Earth-to-Moon shuttle to be used for resupply and personnel change-outs for a permanent lunar settlement. Chemical rockets had proven successful for lifting heavy objects off the ground and sending them to the Moon, but they lacked the efficiency and versatility of the nuclear engines. A chemical rocket could be started only twice, if specially designed, and it could not wait in orbit for lengths of time, waiting to be started. The most advanced chemical rocket made for the Apollo Moon landing program, the Rocketdyne J-2 third stage engine for the *Saturn V* booster, was capable of one restart, and it burned for a maximum of eight minutes and 20 seconds. The NERVA NRX ran for nearly two hours, with 28 restarts.

A final NERVA engine, the XE, was built specifically for spaceflight. For a RIFT launch, the Saturn S-IVB third stage with a J-2 engine was to be changed to a Saturn S-N nuclear-powered stage. The first use of the NERVA XE was planned to be a manned expedition to Mars, to follow the successful Apollo Moon missions. After the Mars landing, a permanent Moon base would use the same type nuclear engine in the shuttle vehicles. The Mars visit was scheduled for 1978. These were bold plans, but the NERVA nuclear engine put them within reach using the technology of the 1970s.

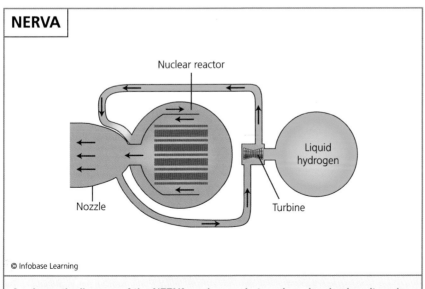

NERVA

Nuclear reactor

Liquid hydrogen

Nozzle

Turbine

© Infobase Learning

A schematic diagram of the NERVA nuclear rocket engine, showing how it works. A small amount of rocket thrust is taken from the exhaust nozzle and used to drive a turbo-pump for fuel delivery.

This ambitious set of programs would require a great deal of money, funded by the U.S. federal budget. The greatest proponents of this expensive space program were President Lyndon B. Johnson (1908–73) and the senior senator from New Mexico, Clinton P. Anderson (1895–1975). Without these men, the funding of such expeditions was not possible. In 1968, Johnson decided not to run for another term as president, and in 1969 Anderson began suffering serious health problems. He was

PROJECT LONGSHOT: A TRIP TO ALPHA CENTAURI

The three-star system of Alpha Centauri, visible in the southern sky Centaurus constellation, is our nearest neighbor in the universe, at a distance of only 4.37 light-years, or 25.7 trillion miles (41.4 trillion km). If we are to send a probing expedition to a star, Alpha Centauri will be a logical first try. Beginning in 1987, NASA and the U.S. Naval Academy in Annapolis, Maryland, conducted a two-year advanced design program called Project Longshot, a one-way trip to Alpha Centauri.

The *Longshot* probe would use a nuclear fission reactor to power it, operating at a continuous 300 kilowatts, in a vehicle weighing 437 tons (396 metric tons). A requirement for this design effort was that it had to use realistic technology as available in the late 1980s. Accelerating to a fast but reasonable 8,333 miles per second (13,411 km/sec), it would take 100 years to get to the star system, where *Longshot* will assume orbit around the second largest of the three stars, Alpha Centauri B. Communication with the probe will be by a 250-kilowatt laser, sending back a tight beam of red light that can blink at a data rate of 1,000 binary bits per second. To send a one-megabyte color image from the Alpha Centauri system will take a little more than 22 hours. A command sent to the orbiting probe will get there in 4.37 years, and the acknowledgment will take another 4.37 years to come back to the Earth receiving station. Operation of the probe will be primarily robotic, given the extreme delays in communication. Data describing the properties of the interstellar medium will be sent to Earth during the flight. The *Longshot* vehicle would be put together in orbit, using the International Space Station as the base of operations. Assembled, it will be approximately 200 feet (60 m) long.

The weak point in the Longshot design is the propulsion system. To reach the high cruising speed requires a rocket engine with a specific impulse of 1×10^6

no longer available to rally the Senate for the NASA exploratory programs. The newly elected president in 1969, Richard M. Nixon (1913–94), promptly cut the NASA expedition budget, ending the Apollo Moon program prematurely and prohibiting any manned trip to Mars. The NERVA nuclear rocket engine program, left with no mission that needed it, was put to sleep in 1972. Anderson resigned from the Senate in 1973.

seconds (1×10^4 kN·s/km), which is higher performance than is achievable by anything that presently exists. To meet this requirement, it was necessary to invent a hypothetical engine using a pulsed fusion micro-explosion drive. Many alternatives were considered in the study, but the needed propulsion characteristics led directly to an engine design similar to that used in the British Daedalus project in 1973. The need for aneutronic fusion, so that all of the fusion product is charged particles and can be directed to the nozzle magnetically, led once again to the use of helium-3. A mixture of helium-3 and deuterium are to be frozen into pellets, directed to the center of the fusion chamber, and hit from all sides by powerful laser beams. The pressure from the collapsing light wave front will compress and heat each pellet to conditions of complete fusion, causing a small thermonuclear explosion. The resulting cloud of hot, expanding plasma pulses the vehicle forward from reaction to the backward directed exhaust. This process is repeated several times per second.

Assuming that 291 tons (264 metric tons) of exceedingly rare helium-3 fusion fuel could be purchased, the main problem with the pulsed fusion engine is achieving a usable reaction using a laser-driven implosion. Such a reaction has been repeatedly attempted over the past four decades, with no reported success. The technical problems of igniting a self-sustaining fusion pulse using anything less than an atomic bomb have proven difficult. The National Ignition Facility in Livermore, California, is presently working on the problems. Early indications are that the laser bank, which is the size of a small town, may be too small.

Project Longshot proved that if we are restricted to technology that can be built and implemented, then we are not at the point where an interstellar expedition is possible. Something new that we have not thought of yet will be necessary for this type of trip. The field is wide open for the next generations to apply new discoveries and knowledge to this very challenging class of problems.

Much talented engineering effort went into developing the NERVA XE engine. It was a mechanically sophisticated engine with very favorable capabilities, and it seemed a shame to abandon it. This engine design was reconsidered for production in the 1980s in a secret program called Project Timberwind, as a rocket for the space-based antiballistic missile weapons developed for the Strategic Defense Initiative. A heavy manned Mars expedition may occur in this century, and it will probably use a NERVA XE engine to get there.

PROJECT PLUTO: THE NUCLEAR CRUISE MISSILE

Fearing a massive nuclear weapons attack from the Soviet Union, the U.S. Air Force and the AEC formulated a plan to counterattack and dissuade such aggressions. In the 1950s, defense from nuclear attack amounted to keeping a squadron of heavy bombers in the air at all times, crossing the world back and forth, waiting for a radioed signal to attack predetermined targets in enemy territory. This strategy was expensive, and it was hard on both the flying machines and the human beings who flew them. Thinking of alternate means, on January 1, 1957, a holiday, the Air Force and AEC contracted the Lawrence Livermore National Laboratory to design, build, and test a new type of weapons delivery system.

The plan was to build a supersonic, low-altitude missile (SLAM). It would enter enemy territory barely aboveground, traveling faster than the speed of sound and peppering the ground with hydrogen bombs. SLAM was to be completely autonomous, requiring no onboard pilot or human crew. It would stay in the air at high altitude for months or years, if necessary, circling around and waiting for a radio command to drop to treetop level and strike. Any ordinary, jet-propelled cruise missile traveling at supersonic speed would run out of fuel in less than an hour, and there was no way it could linger at high altitude for as long as a day. The solution to this daunting problem was clear. The jet engine on board SLAM would have to be powered by a nuclear fission reactor.

The project was code-named Pluto, and to test its nuclear engine a secret laboratory was set up on eight square miles (21 km^2) of desert at Jackass Flats, Nevada, at a cost of $1.2 million. Not revealing its purpose, the facility was named Site 401. The engine, named Tory, was quite simple, with the only moving parts being the reactor control rods. It was a ram jet, taking air in through the front, heating it in an incandescent reactor core, and exhausting it out the back. The vehicle would be traveling at supersonic speed through the thin air at high altitude, and no compressor would be necessary in front of the hot reactor. The engine would be started by lifting Pluto off the ground and into the stratosphere using two solid fuel rockets. After burnout, the rockets were jettisoned, and by that time the missile was traveling at about Mach 3, three times the speed of sound. Start up the reactor, run it up to full power, and the ram jet would kick in, maintaining the high speed indefinitely.

Designing such a missile was enormously challenging. The engine had to run at 2,500°F (1,600°C), only 150°F (66°C) below the ignition temperature of the airframe. The reactor would have to be made of 300,000 pencil-sized fuel rods, made of a special ceramic material built by the Coors Porcelain Company of Golden, Colorado. Another problem was testing a ground-based, static jet engine that was supposed to be flying at a supersonic velocity. For this to work, air had to be forced into the intake funnel at flight speed. To accomplish this, a compressed air tank 25 miles (40 km) long was built of oil well casings, welded together. Giant air compressors were borrowed from the navy's submarine base in Groton, Connecticut, to load the tank with 1 million pounds (450,000 kg) of air. To run the engine, a ton (0.9 metric ton) of air per second had to be forced into the jet intake.

There was concern over the radiation that was certain to be a byproduct of the high-performance nuclear reactor constantly shedding pieces of burned fuel out the exhaust. It was decided to turn on the reactor only when Pluto was over the Pacific Ocean, after the rockets had fallen away in a launch. Just to be safe, Site 401 was equipped with a fallout shelter and two weeks of food for the entire staff, in case the engine exploded and contaminated the test area. A full flight test of Pluto was a special problem. It was a guided missile, not an airplane, so it was not equipped to ever land. It would make too much of a cleanup problem if allowed to crash in the desert, so it was decided to fly it in circles around Kwajalein Island, in the Pacific, and dive it into the ocean upon completion of the test.

The first engine to be tested was Tory-IIA, mounted on a railcar and rolled to the test stand at Site 401. On May 14, 1961, the compressed air valve was opened and the control rods were pulled out by remote control. The reactor power was leveled off at a fraction of its rated power, and the engine had only enough compressed air to run a few seconds, but the test was considered a success. The observers from the AEC were delighted, as the engine had not caught fire as some had predicted.

Exterior view of building number 2201, a disassembly building at the Pluto facility at the Nevada Test Site. It was used for taking apart the reactors after testing for study and improvements. *(Library of Congress)*

Encouraged by the brief performance of Tory-IIA, the Project Pluto engineers devised an improved Tory-IIC and tested it three years later. It ran at full power, 513 megawatts, for five minutes, generating an impressive 35,000 pounds (156,000 N) of thrust. The run time was far from the several months in the product specification, but it was definitely a move in the right direction.

Impressive as it was, other parallel projects looked even more promising, and the disadvantages of the Pluto missile became evident. Although it would fly low, it flew hot, noisy, and radioactive. The engine made a screaming noise at 150 decibels, and you could hear it coming many miles away. It was hardly necessary to carry nuclear weapons, as it would leave a path of destruction on the way to the target with its sound barrier shock wave and heavy radiation contamination. Simply crashing it into the ground would leave a city unlivable with nuclear debris. No one knew what would happen if a Pluto cruise missile ran into a rain shower at Mach 3. It turned out that ballistic missiles using chemical rockets were cheaper, safer to handle, and there was no defense against them.

On July 1, 1964, the project was cancelled for lack of need—it had spent $260 million and employed 450 people for seven and a half years. The topic of a nuclear ramjet was never mentioned again by the air force.

Many of the problems of nuclear propulsion have thus been worked out in advance, should we decide in the future to cast aside the limitations of jet fuel and chemical rockets. The nuclear engines are surely in the future of space travel, and with much refinement nuclear jet engines are not impossible. The next chapter predicts the land-based future of a nuclear society, where oil is no longer burned and coal is left in the ground.

7 Alternate Nuclear Economies

In the past 300 years, industrialized civilization could have evolved along many different tracks, any of which involve energy production and consumption as a basis. After these centuries of progress and constant adjustment and gradual change, we have settled into an energy economy that depends almost exclusively on oil products for air, ground, and water transportation. Home heating and lighting, as well as industrial motive force, are almost entirely accomplished using distributed electrical power, with a small percentage carried by natural gas distribution. The source of our distributed electricity is mixed. Most of it is produced by conversion from coal and natural gas, one-fifth of it is produced by nuclear fission, and small amounts are produced by wind, sunlight, and hydro-dams. Coal is a limited resource dug out of the ground, as is oil and natural gas. No new coal, natural gas, or oil is being made and stored in the ground, so these energy resources are considered nonrenewable. When all the coal, oil, and gas are gone, there will be none to replace it. Coal reserves are large enough that we can keep on burning it at this rate for hundreds of years. Oil, however, is becoming scarce, and an alternate fuel for transportation, at least, must be found and engineered.

Electricity is our most versatile form of energy, because there are so many alternate ways of producing it. It can be transported using copper

116

wires, whereas transportation of liquid fuel, such as gasoline, requires buried steel pipes and tanker trucks. Some alternative transportation systems not involving gasoline or diesel fuel have already undergone some development and use, such as electric cars and trains. Electric passenger planes and 18-wheeled trucks are more difficult to imagine, but the lack of oil resources may reach the point where we will have to conserve it where possible. This will eventually mean that passenger cars and trains, at least, will have to be powered by electricity or some other alternate fuel, saving the petroleum products for special purposes.

Nuclear power as it now exists was designed for the economy of the 1960s, when there was a concentrated push for applying fission reactors to consumer needs for electricity. The typical nuclear plant is a large installation, near a river, producing a base load of bulk electricity at a rate of about 2 billion watts per plant. There are other uses for the heat of nuclear fission in addition to making steam and turning turbogenerators. This same heat source can be used to convert water into hydrogen fuel, without the inefficient middle step of first converting the power to electricity. Hydrogen can take the place of petroleum in many applications, including personal transportation. Run the country on hydrogen produced by nuclear power and the dependence on dwindling oil stocks falls away. Nuclear fission can also be used for desalination, making fresh drinking water and irrigation water from seawater, and making otherwise dried-out land productive and life supporting.

These are fine applications for nuclear power, but the fuel used to make nuclear heat is no more renewable than oil or coal. However, there is far more energy available from uranium-235 than is available from fossil fuels, and there is a bonus. Less than 1 percent of all the uranium mined is usable fuel, uranium-235. The rest is uranium-238, which cannot fission and produce heat. However, using a breeder reactor, the otherwise worthless uranium-238 can be converted to fissile plutonium-239. If we have enough uranium-235 to last 1,000 years making energy, then with breeder reactors and a plutonium energy economy, we can go 100,000 years.

All of the nuclear power hopes are not based on uranium. There is another reactor fuel, thorium. There is at least three times more thorium that can be mined than there is uranium, and 100 percent of it can be used as fuel. By exploiting the alternate nuclear fuels thorium and plutonium, it is possible to produce electricity, hydrogen gas, and freshwater for the foreseeable future without burning a single bucket of coal.

This chapter explores these alternate economies, from hydrogen production and use to water desalination. The practical uses of the alternate fuels are discussed, and the sidebar mentions the special handling techniques necessary for plutonium, which in 30 years could be producing electricity for your fluorescent bulbs.

THE SULFUR-IODINE CYCLE AND A HYDROGEN ECONOMY

In the 1970s, General Atomics, a nuclear research center in San Diego, decided it was time to branch out. Instead of assuming that nuclear power would solve all energy problems, they saw that an alternative to gasoline would eventually be needed for personal vehicles, which were too small to contain a fission reactor. A simple, nonpolluting alternative to gasoline is hydrogen gas. There are multiple ways to use it for small-scale propulsion, and burning it in air produces no carbon dioxide. The product of burning hydrogen in air is simply water, and it is a very efficient chemical reaction.

In the hydrocarbon economy, such as now exists in the industrial world, petroleum is the main fuel for transportation, and coal is the main fuel for energy production. Petroleum and coal are limited resources, and they will become less abundant and more costly as the underground reserves dwindle. Combustion products in air are mainly carbon dioxide and water, with partially burned components such as carbon monoxide and soot in smaller concentrations. Traces of sulfur impurities in these fuels oxidize and pollute the combustion exhaust. Nitrogen in the air can oxidize in the heat and pressure of combustion, making further undesirable pollutants in significant concentrations.

There is no hydrogen mine as such, and hydrogen gas is not available in nature. However, hydrogen gas is made from naturally occurring substances for use in industry, such as the manufacture of ammonia fertilizer. Worldwide, about 55 million tons (50 million metric tons) of hydrogen are produced yearly. It would take 187 million tons (170 million metric tons) of petroleum to make the same energy as burning the world's yearly hydrogen product. Another use for this industrial hydrogen is hydrocracking, a process by which low-quality hydrocarbon resources, such as tar and even coal, can be converted into gasoline. It is estimated that 41.6 million tons (37.7 million metric tons) of hydrogen per year could be used to convert coal to gasoline, and that this process would end the U.S. dependence on foreign oil imports.

Current hydrogen production is by steam reforming of oil, natural gas, or coal. The exhaust from this process is carbon dioxide, and obviously this method misses the goals of a hydrogen economy by using nonreplenishing resources, polluting the air, and requiring even more hydrocarbon to heat water into steam.

If there were an efficient, nonpolluting way of making hydrogen that did not use up oil, gas, and coal reserves then it could be used to power cars using two possible methods. One is to use hydrogen as if it were gasoline, injecting it into a spark-ignition, internal combustion engine, slightly modified to use gaseous hydrogen instead of liquid petroleum. The first use of hydrogen gas in an internal combustion engine was in 1807 by the French politician and inventor François Isaac de Rivaz (1752–1828). Living in Switzerland after retirement from the French army, he installed his newly built engine in a four-wheeled vehicle in 1808. He named it "automobile." Attempts to sell his novel invention were unsuccessful. Further exploitation of hydrogen as an internal combustion fuel would wait until the Mazda Motor Corporation of Hiroshima, Japan, built an experimental Wankel automobile engine using direct hydrogen injection. Exactly 200 years after de Rivaz built his automobile, a hydrogen-powered forklift was demonstrated in Hannover, Germany, using a modified Linde X39 diesel engine.

Hydrogen has been slow to be implemented as an internal combustion fuel partly because in the heat of burning it reacts with nitrogen in the air to make such pollutants as nitric acid and hydrogen cyanide. The requirements for the next fuel to be used in cars are that it must be renewable, not from a foreign source, and it must not pollute the air.

A second method to use hydrogen to power a car, the electric fuel cell, is even more promising than using converted gasoline and diesel engines. The fuel cell burns hydrogen in air, as does an internal combustion engine, but it does so at a controlled rate, using two catalysts. The exhaust is water vapor. The fuel cell has no moving parts, there is nothing to wear out, and it produces 0.6 to 0.7 volts of direct current electricity per cell. Like batteries, fuel cells can be connected in series to give higher voltages. The efficiency of a fuel cell is about 50 percent, with half of the energy used to produce electricity and half wasted as heat. An automobile engine or a nuclear reactor are about 30 percent efficient.

A fuel cell consists of a sheet of electrolyte between an anode catalyst plate and a cathode catalyst plate. Hydrogen gas is piped to the anode plate, and the cathode plate is exposed to air. The anode catalyst breaks the hydrogen down into positive ions and electrons, and the electrons flow from anode to cathode. In the cathode, the hydrogen ions are neutralized by combining with the oxygen, and the water vapor exhaust pours from this plate.

This would seem an ideal way to make electric cars that do not depend on inefficient batteries to store energy. A major stumbling block to adopting electric automobiles for transportation is the limited amount of energy that can be stored in batteries. By storing electrical energy instead as hydrogen in a tank, the range and utility of an electric vehicle becomes closer to what we are accustomed to with gasoline-powered vehicles. Hydrogen filling stations have been designed and demonstrated, filling the hydrogen tank on a fuel cell car as one would fill a gas tank. A possible disadvantage to using fuel cells in electric passenger cars is the cost of

The Chevy Equinox fuel cell vehicle in 2008 at the EVS23 Conference in Anaheim, California (*DOE Photo*)

A driver at a hydrogen pump in Los Angeles, California, refueling a zero-emission Chevy Equinox fuel cell automobile. The vehicle can travel 150 miles (241.4 km) on a single fill-up. An electric plug-in cable controls the refilling process.   *(Spencer Grant/Alamy)*

the catalysts. The cathode plate is made of nickel, which is not excessively expensive, but the anode is made of platinum, and this could drive up the cost. Cheaper alternatives are being sought.

The first fuel cell–powered electric vehicle was a D12 farm tractor, demonstrated by the Allis-Chalmers Manufacturing Company in 1958. In 1960, Allis-Chalmers built the first fuel cell–powered forklift, in 1964, a one-man underwater research vehicle, and, in 1965, the first fuel cell–powered golf cart. In 1966, the first attempt at a fuel cell-powered automobile, the Electrovan, was built by General Motors Corporation. It was able to travel at up to 70 miles per hour (113 km/hr) for as long as 30 seconds. Development of the fuel cell–powered electric car continues.

What hydrogen needs to make it a viable alternative to petroleum is a manufacturing method that does not involve petroleum. Such methods have been developed, and both the hot electrolysis and the sulfur-iodine

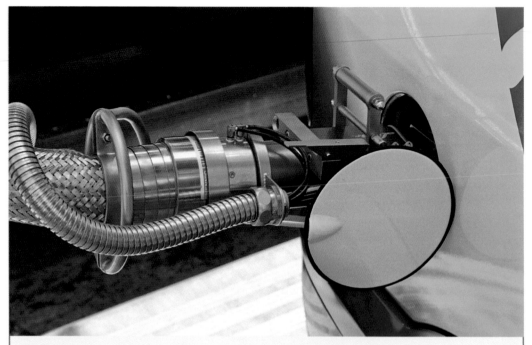

A hydrogen nozzle refueling an Opel Hydrogen 3 fuel cell car at the Michelin Challenge Bibendum 2006 in Paris, France   *(Dieter Wanke/Alamy)*

cycle schemes work well using the Generation IV very high temperature gas reactors (VHTRs). For the hydrolysis method, two electrodes are immersed in water. Power is applied to the electrodes, and the water breaks into its two components, hydrogen and oxygen. Oxygen gas streams off the anode, and hydrogen bubbles from the cathode. Heating the water reduces the electrical power necessary to break the chemical bond between the hydrogen and the oxygen.

The VHTR heats helium to 1,740°F (950°C), which is used to drive a gas turbine making electricity, as at a conventional nuclear power plant. A second helium loop heats water for hydrogen production at the same plant. A percentage of electrical power is tapped off the generator and operates a bank of hydrolysis units, which are fed freshwater from a heat exchanger taking energy from the second helium loop. Hydrogen gas off the hydrolysis cell cathodes is compressed into tanks for distribution, and the oxygen is vented into the atmosphere. During off-peak hours, such as at night when people are asleep and not using electricity, a larger percentage of the electrical power is diverted into the hydrogen production building.

The *sulfur-iodine cycle (S-I cycle)* is also ideally suited for operation with a VHTR. Its efficiency is around 50 percent, meaning that about half of the energy required to make hydrogen using this method is recoverable by reoxidizing the hydrogen. Operating at a practical temperature

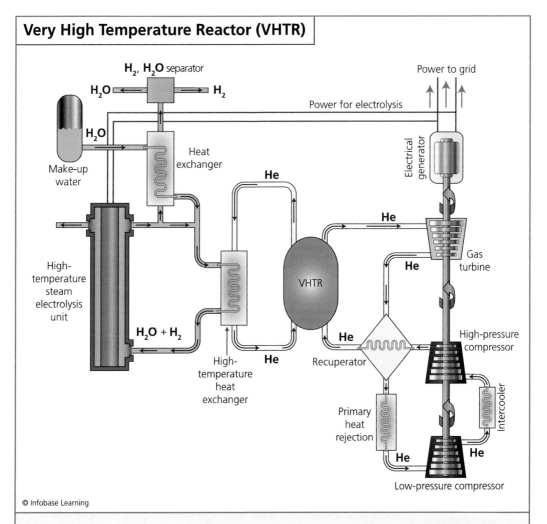

Very High Temperature Reactor (VHTR)

© Infobase Learning

A very high temperature gas reactor (VHTR) is shown in a dual-purpose mode in this schematic. The reactor has two helium coolant loops. One directly drives a turbogenerator to make electricity, and the other heats up water for high-efficiency electrolysis, making hydrogen out of water.

of 212°F (100°C), hydrolysis is about 41 percent efficient. The S-I cycle is a three-step chemical cycle requiring external heat, which is supplied by a hot helium gas loop in a VHTR.

The three hydrogen production reactions, each of which requires a separate applied temperature, are as follows:

✳ Reaction 1

$$I_2 + SO_2 + 2H_2O \rightarrow 2HI + H_2SO_4$$

Iodine plus sulfur dioxide plus water are combined at 248°F (120°C), yielding hydrogen iodide and sulfuric acid. The water is introduced from outside the cycle, and is consumed.

✳ Reaction 2

$$2H_2SO_4 \rightarrow 2SO_2 + 2H_2O + O_2$$

Sulfuric acid is heated to 1,526°F (830°C) and decomposes into sulfur dioxide, water, and oxygen gas.

✳ Reaction 3

$$2HI \rightarrow I_2 + H_2$$

Hydrogen iodide is heated to 842°F (450°C) and decomposes into iodine and hydrogen.

The net reaction is $2H_2O \rightarrow 2H_2 + O_2$, or water decomposes into hydrogen gas and oxygen gas. The iodine and sulfur dioxide by-products are cycled back into the system and never have to be replenished. The only inputs are water and heat to achieve the reaction temperatures. The heat is supplied by a reactor running at power, and the water is supplied by a river or lake. The exhaust from the process is oxygen gas.

Successful experiments have been conducted in Japan with the intent to use this process with a Generation IV reactor to replace a dependence on gasoline for automobile fuel. There is also work being conducted under an International Nuclear Energy Research Initiative jointly with the French CEA, General Atomics, and Sandia National Laboratories. Addi-

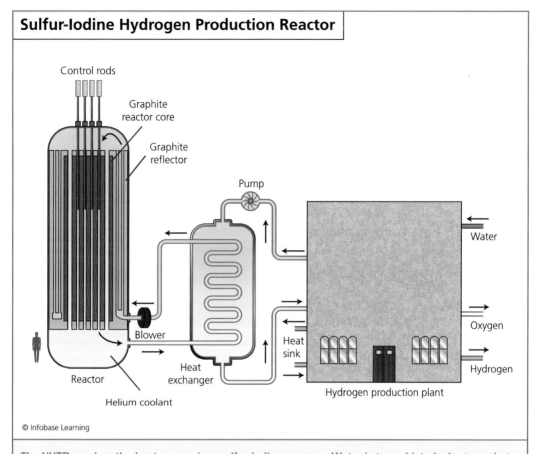

Sulfur-Iodine Hydrogen Production Reactor

Control rods

Graphite reactor core

Graphite reflector

Pump

Water

Blower

Heat sink

Oxygen

Reactor

Heat exchanger

Hydrogen

Helium coolant

Hydrogen production plant

© Infobase Learning

The VHTR used as the heat source in a sulfur-iodine process. Water is turned into hydrogen using only the heat from the reactor. Burning the hydrogen in a car does not pollute the air and nor does the production of hydrogen fuel.

tional research into the use of the S-I cycle is being conducted in Canada, Korea, Italy, and the Idaho National Laboratory.

THE FAST BREEDER REACTOR AND A PLUTONIUM ECONOMY

At this point in the implementation of nuclear power, with 14 percent of the world's electrical power generated by nuclear means, the demand for uranium sets the price for raw uranium oxide. It reached a low of $7 per pound in 2001, and it has been as high as $113 per pound in 1977. The average price is between $55 and $70 per pound. This may change in the future, as countries such as China and India press for industrial upgrades, and their electrical demands grow rapidly. Once plentiful reserves of uranium will decrease, and the price of reactor fuel may become significant. Uranium-235 is not a renewable resource, and eventually it will run out.

Although this is not an immediate concern, it will be in the future. While the largest reserves that are practical for mining are in Canada

SPECIAL CAUTIONS WHEN WORKING WITH PLUTONIUM

Plutonium is a silvery metal, heavier than gold, extremely rare in nature, and dangerous to handle. The fissile properties of a special plutonium isotope, plutonium-239, make it a useful material for power production, but as an industrial material it is as hazardous as radium, and special precautions are necessary.

There are three uses for plutonium. Plutonium-238 has a short half-life of 87.8 years and it decays by alpha emission. A small pellet of this very active isotope will glow red from the heat it gives off, and therefore pieces of plutonium-238 are used as highly reliable, solid-state sources of heat in unmanned spacecraft sent on multidecade expeditions to explore outer planets in the solar system. Plutonium-239 is used as reactor fuel in power plants, but it is also the active ingredient in nuclear weapons. These isotopes are manufactured from uranium-238 stock using specially built production reactors.

Plutonium is pyrophoric, meaning that a block of it exposed to damp air will react quickly and can burst into flame. A burning chunk of plutonium gives off dense white smoke, and if it is not flaming it at least resembles a glowing ember. The smoke will quickly contaminate any object or person it touches with dangerously radioactive material. The ball of plutonium used in nuclear weapons typically weighs 11 pounds (5 kg). Even with a half-life of 24,390 years, from its alpha decay alone such a mass of plutonium-239 produces 9.8 watts of power, and it makes a perfect hand warmer for technicians assembling atomic bombs in the cold. These bomb components are plated with nickel to prevent the plutonium from being exposed to air and catching fire. One must be careful not to scratch the plating, but the alpha particles from the nuclear decay are quite harmless in this situation. Only introduction into the metabolism is dangerous, for the same reason radium was dangerous when it was used to and Australia, most of the Earth's uranium is dissolved in the oceans. Uranium oxide is soluble in water, and over billions of years of rain, it has been dissolved out of rocks and washed out to sea. There is an estimated 8.8 trillion pounds (4 billion metric tons) of it in the oceans. Research is underway to develop a filter that will capture the uranium in seawater using a passive collection strategy. Indications are that large-scale ura-

make glow-in-the-dark watch dials. As with radium, ingested plutonium will find its way into the bones and the alpha decay will cause irreversible destruction to tissues.

Plutonium is considered to be chemically toxic as well as radiologically hazardous. It is less dangerous than the element arsenic or the compound cyanide. It is about as toxic as caffeine.

Although it is considered one of the most dangerous substances on Earth, no human is known to have died because of inhaling or ingesting plutonium, even though many people have measurable quantities of it in their bodies. Anyone living downwind from the Nevada atomic bomb test area in the 1950s breathed a lot of plutonium, as did the survivors of the Nagasaki nuclear attack and the population of Japan in 1945. Predictably, plutonium has a metallic taste. In 2005, there were between 50 and 100 implanted plutonium-powered heart pacemakers still functioning in living people.

Laboratory animal studies in 1945 showed that a few milligrams of plutonium is a lethal dose, but definitive safety standards could be determined only by implanting it into human subjects. In a study widely criticized for an obvious breach of medical ethics, 18 patients, thought to have less than 10 years to live, were injected with solutions of plutonium without their knowledge or consent from October 1945 though July 1946. Although no patients died directly from plutonium or its associated radiation, the studies were not inconsequential.

The final caution for working with plutonium is the accident of placing two subcritical masses of it close together and making them nuclear reactors, spewing energy, radiation, and fission products in an uncontrolled manner with no shielding. Although such an accident would seem unlikely, it happened twice at the Los Alamos National Laboratory in 1945, soon after the end of World War II. Two working scientists, very familiar with the process of self-sustained nuclear fission and the fissile properties of plutonium, managed to accidentally touch two hemispheres of the metal together in two separate accidents. Both men died of acute radiation poisoning from the resulting uncontrolled nuclear fission.

nium extraction from water will cost about $130 per pound. When mines play out, further deposits are difficult to identify, and the cost rises above this estimate, seawater extraction plants will be built.

This finite source of reactor fuel may not curtail the use of nuclear power for several centuries, but it still makes uranium fission a temporary solution to the needs of civilization. Fission reactors, in fact, make inefficient use of uranium. Less than 1 percent of it is the fissile uranium-235 isotope, and the rest is the inert isotope uranium-238. Only about 5 percent of the useful uranium is fissioned, and the rest is thrown away with the waste products. When uranium becomes scarce, the power industry will not be able to afford to discard usable fuel, and spent fuel rods will be reprocessed to extract and repurify the uranium-235.

The vast majority of the uranium, more than 99 percent, is not fissile. However, this isotope, uranium-238, will capture stray neutrons in the fission process and activate into uranium-239. It has a half-life of only 23.5 minutes, and it undergoes a beta-minus decay into neptunium-239, which decays in similar fashion into plutonium-239. The neptunium has a half-life of only 2.35 days, but the plutonium-239 will remain in place far longer, with a half-life of 24,390 years. Plutonium-239 is fissile, and it makes a serviceable reactor fuel. A reactor fueled with uranium therefore makes extra fuel as it makes power, converting otherwise worthless uranium-238 into fissile plutonium-239 with wasted neutrons. Near the end of a fuel cycle, which is typically three years, a pressurized water reactor (PWR) or a boiling water reactor (BWR) is burning derived plutonium as well as uranium. In spent reactor fuel, about 1 percent of the waste material is plutonium. Worldwide, almost 110 tons (100 metric tons) of plutonium are produced in spent fuel every year, even though most of it is fissioned away during the fuel cycles.

In the 1980s, the United States and the Soviet Union mutually agreed to cut back the number of stockpiled nuclear weapons, reducing the threat of nuclear warfare. Thousands of bombs and warheads were dismantled, and this left many tons of plutonium-239 atomic bomb cores unused. Rather than store or bury this vast amount of fissile material, it was decided to use it as reactor fuel. In small batches, plutonium was oxidized, mixed with uranium oxide, and formed into reactor fuel pellets. It is called MOX, or mixed oxide fuel. Only about 3 percent of the MOX is plutonium, and at this rate it may take hundreds of years to get rid of all the surplus plutonium-239. A standard reactor can fission MOX fuel without modifications, but only if the fuel load is less than half MOX and half standard, 3 percent uranium oxide. For using more than half MOX, the reactor must be specially built, with additional control rods. Pure MOX fuel tends to run hotter because of a slightly lower thermal conductivity.

Although the Palo Verde Nuclear Generating Station in Arizona was built to run pure MOX fuel, and any other reactor in the United States could run 30 percent MOX, no plant is licensed to use it yet. The Tennessee Valley Authority and Duke Energy have applied for licenses, and a MOX fuel plant is being built at the Savannah River Site in South Carolina, scheduled for startup in 2016. About 50 reactors in Europe have been licensed to use MOX fuel, and Japan has loaded 18 reactors with one-third MOX since 2010.

Favorable experience with using MOX fuel will prepare the power industry for a further step in which reactors are specifically built to run pure plutonium-239 fuel. Plutonium fuel is less expensive to make than uranium fuel, because it requires no isotope enrichment. Although it contains isotope contaminants such as plutonium-240, there is no need to separate out minor concentrations. A further step in the development of plutonium reactor fuel will be to use fast neutrons for fission instead of thermalized neutrons, slowed down for optimum fissioning. Liquid metal will be used for cooling instead of water, to avoid slowing down the neutrons, and the excess fast neutrons leaking from the fission process will be used to efficiently convert uranium-238 into plutonium-239. The result will be breeder reactors that make more fuel by conversion than they burn, using the vast stockpile of unused uranium-238.

Experimental breeder reactors have been built, run, and torn down many times in the continuing effort to understand this complicated technology. In Germany, construction of the SNR-300 full-scale fast breeder reactor in Kalkar, Westphalia, was completed in 1986. In 1991, the SNR-300 was officially cancelled. In 1995, a Dutch developer transformed it into an amusement park, now named *Wunderland Kalkar*. Presently, the only active fast breeder reactor in the world is the BN-600, built in 1986. It is still running and producing 600 megawatts of electricity plus excess fuel in Zarechny, Russia. Breeder reactors are in the plans for the Generation IV power plants. The use of breeder reactors and an energy economy that runs on plutonium may be the final development of using fission to make power in the far future.

THE THORIUM ECONOMY

India is looking into the future with detailed plans for energy expansion and independence. There is now a three-stage nuclear power program underway in India, and the final step is to make use of energy locked in the monazite sands of Kerala and Orissa, India. Monazite contains thorium phosphate, and thorium can be used as a reactor fuel.

Stage I of the Indian plan is to build pressurized heavy water reactors, based loosely on the venerable CANDU design, and using natural uranium as fuel. These reactors will produce both electrical power and plutonium-239, which will fuel the Stage II reactors. Stage II is to build fast breeder reactors using the plutonium-239, and they will produce electrical power and still more plutonium-239 using the depleted fuel from Stage I as breeding stock. This will keep India energy independent until the uranium runs out. The first plutonium breeder reactor is to be commissioned in 2012. Finally, in Stage III, the advanced heavy water reactors will be built, first using a MOX fuel consisting of plutonium oxide and thorium oxide. These reactors will gradually switch to using only thorium oxide fuel, and India will retain its fast-running economy, independent of any outside energy sources.

By first using a mixture of oxide fuels, the Indian thorium reactors will sidestep a disadvantage to thorium reactor fuel. Thorium-232 itself is not fissile, but it activates by neutron capture into thorium-233, which quickly beta decays into protactinium-233. The protactinium then beta decays with a 27-day half-life into fissile uranium-233. This 27-day delay for half of the protactinium to become usable fuel makes it impossible to cold start a reactor fueled solely with thorium. In the Indian plan, there is enough premade uranium-233 in the fuel to form a critical mass, and the long-term fuel is then converted at leisure using the thorium oxide component of the fuel. This process can be started initially without any uranium-233 by using plutonium from the fast breeder program as starter fuel. This plan is well thought out and will result in the vast power system that will be necessary to drive the economy that India is planning to achieve.

There is another way under development to use thorium as an energy source. It is the subcritical reactor or the accelerator-driven system (ADS). The original concept of nuclear power was to have a mass of fissile material kept at criticality, a state in which fission-causing neutrons are kept in perfect balance, with all the neutrons lost by capture or leakage made up for with neutrons produced by the fissioning fuel. The product of the controlled fissions is energy in the form of heat.

It is not necessary for a mass of fissile material to be critical to produce energy. In a subcritical assembly of fuel, an incoming neutron causes a fission. This fission ejects two more neutrons, which cause two more fissions, and so forth. However, more neutrons are lost than are necessary to continue the reaction, and the energetic excursion is just a transient surge of power. That one neutron was still productive, making billions of energy-producing fissions at the cost of just one neutron. The phenomenon is called neutron multiplication. If there was an external source of neutrons driving such a system, it would have advantages. There is no danger of a runaway reaction. When the neutron source is turned off, the energy production stops. It would be a nuclear power source that could be turned on and off with a light switch. Also, it will burn anything that can respond to a neutron, with no requirement that each reaction must be a neutron-producing fission. This means that it can burn the transuranic components of nuclear waste from a conventional reactor. Not only can it use otherwise buried material for fuel, it also reduces nuclear waste in the process. While the heavy transuranic components in nuclear waste can be radioactive for hundreds of thousands of years, the lighter components last only hundreds of years as radioactive decay eventually renders them inert.

A method under development in Europe and Japan to power a subcritical reactor is to use an electrically driven proton accelerator. This ADS uses an industrial high-energy proton source to direct a beam at a thorium target in the reactor. When hit by a high-speed proton, a heavy thorium atom spallates into an average of 20 neutrons. These neutrons go on to initiate subcritical fission in nearby uranium-233 atoms, and net power is produced. If a pure thorium fuel loading is used, then it will take weeks of shooting proton beams into the reactor before enough uranium-233 is made by neutron capture to begin power production. An ADS reactor can start using MOX fuel, just as a conventional reactor can.

This precisely controlled method of power production may be incorporated into the Generation IV reactor concepts as a small-scale energy source, a waste management function, or as a hybrid connection to a larger system of fast breeder reactors.

NUCLEAR-POWERED WATER DESALINATION

Desalination refers to any process by which seawater can be converted into potable freshwater. In countries without free-flowing rivers, lakes, or significant rainfall, the only way to sustain human habitation and agriculture is to pump water out of the ocean and desalinate it. As climate changes dry up otherwise livable sections of the Earth and populations increase, desalination becomes critically important.

There are many ways to desalinate water, from electrodialysis to reverse osmosis. Several methods are presently under investigation, including solar and geothermal desalinization, but on a large, industrial scale nothing outperforms distillation, or evaporating the water using applied heat and then condensing it back to liquid form. Specifically, multistage flash distillation is the most widely used method with seawater. In a flash distiller, steam is used to heat incoming salt water and drain it into a large, horizontally mounted tank, divided into five sections. The hot water drains from section to section, becoming more and more salty as water evaporates away, and finally draining from the fifth chamber to be returned to the sea. Salt is kept in solution and does not encrust the

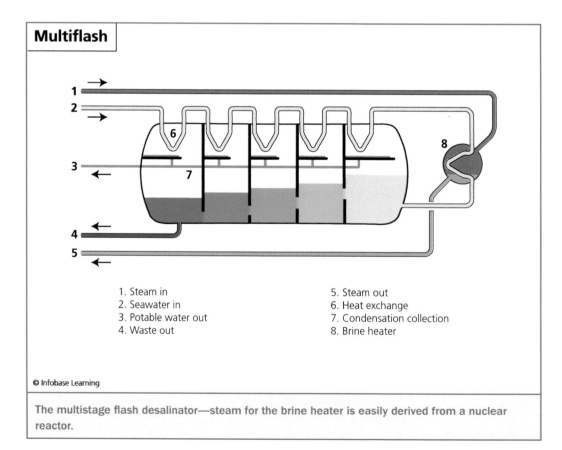

Multiflash

1. Steam in
2. Seawater in
3. Potable water out
4. Waste out
5. Steam out
6. Heat exchange
7. Condensation collection
8. Brine heater

© Infobase Learning

The multistage flash desalinator—steam for the brine heater is easily derived from a nuclear reactor.

inside of the tank. To recover the vapor as liquid water, cooling pipes are located in the top of each of the five tank segments, and the incoming salt water acts as a coolant, running through the pipes on the way to the steam heater. External power is required to supply the steam and to run the saltwater pump.

Worldwide, 13,080 desalination plants are presently at work, producing more than 12 billion gallons (45 million m^3) of freshwater daily, and approximately 85 percent of these plants use flash distillation. The largest facility in the world is the Jebel Ali Desalination Plant, Phase 2, in the United Arab Emirates. It is capable of processing 216 million gallons (822 thousand m^3) of water per day using petroleum to supply heat and power.

In the interest of saving gas, oil, and coal for other uses, it is possible to use nuclear power to make the steam and electricity used in flash distillation. Dual-purpose reactors are presently used in India, Japan, and Russia to produce both electrical power and freshwater. In Japan alone, eight reactors are coupled to desalination plants. There are no fixed plants in the United States to derive freshwater from nuclear power, but a single aircraft carrier in the U.S. Navy can use nuclear power to desalinate 400 thousand gallons (1.5 thousand m^3) of water per day. In 2011, an aircraft carrier supplied freshwater to cool the Fukushima I Nuclear Power Plant reactors and fuel storage pools after heavy damage from the Tōhoku earthquake and tsunami.

Desalination is a practical example of how heat from nuclear fission will be used in the future, reserving dwindling supplies of petroleum for applications with few alternatives, such as aviation fuel. As the number and the extent of such installations grow, the need for the last step in nuclear power production, waste disposal, will increase as well. The next chapter discusses new and old ways to deal with disposable fission products, in step with the Generation IV reactor developments.

8 Advanced Concepts of Long-Term Nuclear Waste Disposal

The process of nuclear fission produces radioactive by-products. As a rule, nuclides that are neutron-heavy, having more neutrons in the nucleus than are necessary for stability, tend to decay toward a more stable configuration, having fewer neutrons per proton in the nucleus. Furthermore, the heavier a nucleus is, the more lopsided its stable configuration is, having more neutrons than protons. Helium-4, a perfectly stable, nonradioactive nuclide, is very light, and it has exactly two protons and two neutrons in its nucleus. Lead-208, which is also stable, is heavy, having 82 protons in its nucleus. To balance the number of protons in a nonradioactive configuration requires 126 neutrons. Lead is neutron heavy, and one more neutron, making it lead-209, makes it beta decay with a 3.32-hour half-life.

Uranium-235 has 92 protons and it is not exactly stable, but with a 700-million-year half-life, it is only slightly radioactive. The half-life of a nuclide is an indication of its degree of radioactivity. Short half-lives mean quick disintegrations and much activity, whereas long half-lives mean slow radiation release. Uranium-235 has an impressive 143 neutrons. It usually loses two or three in the fission process, and this leaves 140 neutrons to be divided between the two large fragments leftover from the energy-producing fission. If fission were symmetrical, then each product would have 46 protons and 70 neutrons, but fission is not symmetrical. One resulting nuclide always turns out heavier than the other one.

A typical product of uranium-235 fission is ruthenium-106, having 44 protons in its nucleus. It has a 369-day half-life, decaying by the beta-minus process up to rhodium-106, with 45 protons. In its beta-minus decay, the excess neutron in the ruthenium becomes a proton and ejects an electron from the nucleus. Rhodium-106 has a 29.9-second half-life and decays similarly by beta-minus into stable palladium-106.

Nuclear decay is an exponential process. Half of the radioactivity is gone in one half-life. Half of what is left is gone after another half-life, leaving one-quarter of the original activity, then half of that goes away in another half-life, and so on. After 10 half-lives, the activity has gone down by 1,024, and the original nuclide is generally considered gone by then. A radioactive substance with an eight-day half-life is gone after 80 days. In theory, there is always a very small residue of the original nuclide left.

This waste component of the fission of uranium-235, ruthenium-106, will thus decay to stable, nonradioactive palladium in 10 years. As is the case of most fission waste, the decay is not a simple single-step process, and each step has its own half-life. In this case, the second decay half-life is too short to be of consequence.

There are 23 principal fission products from nuclear power production. The most prevalent, in 6.8 percent of fission debris, is cesium-133. The most rare is metastable cadmium-113, only appearing in 0.0003 percent of fissions. Two fission products, sumerium-149 and gadolinium-157, are not radioactive. In addition to fission products, nuclear waste also includes radioactive transuranic nuclides, or nuclides made by the original fuel components capturing neutrons. Uranium-238, for example, activates by neutron capture into plutonium-239, and plutonium-239 activates into plutonium-240.

Half-lives of fission products range from iodine-131, at eight days, to iodine-129, at 15.7 million years. The exponential decay curves of all the products, plus all the subsequent radioactive decay curves products resulting from first, second, and later decays, combine into one long decay curve. While most of the radioactivity left over from fission is gone 40 minutes after a reactor has shut down, it can take 100,000 years for it all to decay down into mere background radiation.

A 1 billion watt power reactor produces about 30 tons (27 metric tons) of spent fuel per year. Of that, only 3 percent is burnable fuel. Only 5 percent of the burnable fuel is turned into fission products in a fuel cycle, which lasts about three years. The rest of the fuel is thrown away with the waste. Worldwide, about 13,000 tons (12,000 metric tons) of spent reactor fuel are produced every year.

Radioactive decay produces penetrating gamma rays as a by-product of the alpha and beta disintegrations, and the energy released from transmutations of elements manifests as heat. The heat production rate goes down with the radioactivity, in a similar exponential curve. In the months following extraction from a working reactor, spent fuel must be kept in a cooling bath of circulating water. Once the heat and radiation have decayed to the point that active water cooling is no longer necessary, the used reactor fuel assemblies are transferred into concrete and metal casks. Here, they wait, stored on the premises of the nuclear power plant and cooled only by air and contact with the ground. The only remaining problem is finding somewhere to put these casks where they can sit quietly and undisturbed for the next 100,000 years.

Many schemes to get rid of radioactive waste have been studied. An obvious thing to do with it is to lift it into outer space and either park it in a solar orbit between the Earth and Venus, or just push it toward the Sun. An obvious drawback would be a failed rocket launch, in which a load of radioactive waste could be blown into the atmosphere as the liftoff vehicle disintegrates. Just one such accident would negate the entire purpose of getting it off the planet. There are also economic drawbacks, as interplanetary rocketry is extremely expensive.

Fission waste could be chemically processed to separate the particularly dangerous transuranic elements, such as plutonium, from the material and subject it to neutron bombardment. Each radioactive nuclide in the waste would then be activated either into a quickly decaying version with a short half-life or a stable version. This process has been a suggested auxiliary function for a fusion reactor, making use of otherwise wasted neutrons produced by the fusion reaction. Unfortunately, self-sustaining fusion power has yet to be achieved, and it may be available only later in this century.

People in the nuclear industry think disposal at the South Pole would seem reasonable. Sealed canisters of fission products could be dumped on the ground in Antarctica, near the South Pole, where nothing lives and perpetual ice covers everything. The heat from radioactive decay would cause the canisters to fall slowly through several thousand feet of solid ice, eventually hitting the rocky ground underneath. This method seems foolproof, but it is not practical because the United States signed the Antarctica Treaty on December 1, 1960. This treaty, signed by 49 countries, states specifically that "the disposal of radioactive waste material shall be prohibited."

An uninhabited island in the Pacific Ocean could be used, and the United States and Japan jointly funded a $3 million study of burying fission products on an island south of Hawaii. A protest in 1979 by the South Pacific Forum made this plan unusable. Disposal in the ocean itself has also been considered. Waste packed into bullet-shaped canisters could be dropped into the deeper parts of the Pacific Ocean. Each canister would fall thousands of feet through the water and hit the clay ocean floor with enough speed to bury itself 100 feet (30 m) down. Clay would eventually cover over the hole, and a canister of radioactive waste would be sealed forever under clay and water shielding. The Seabed Treaty of February 11, 1971, signed by 87 countries, forbids this mode of fission waste disposal.

THE YUCCA MOUNTAIN REPOSITORY

In 1957, just as the United States began to experiment with commercial nuclear power, the National Academy of Sciences held an international meeting to discuss and plan for the safe disposal of fission wastes. The resulting report recommended burying it, deep enough to make it unavailable for human contact. Furthermore, it should be placed so as not to contaminate drinking water. An ideal resting place would be in a deep salt mine, as water has never been there. If water had ever invaded such a geological structure, it would no longer be a salt deposit. It would be a cave, as the salt would have dissolved out.

The first implementation of this suggestion, beginning in 1967, was in the Asse II pit, an abandoned salt mine in the Asse mountain range in Wolfenbüttel in Lower Saxony, Germany. The facility was used experimentally until 1978, and 125,787 drums of low-level and 1,293 barrels of medium-level radioactive waste were stored there. It was an interesting test of the concept, but the old pit had been designed for mining salt and not for storing things bound into metal drums. There were no columns and struts to stabilize the chambers, and deformation, fractures, and water breaches in the salt were considered likely. The facility is being monitored for radiation leakage, but no further waste deposition is allowed.

In 1974, the Department of Energy (DOE) began planning for a nuclear waste repository in the United States, named the *Waste Isolation Pilot Plant (WIPP)*. After 20 years of research, the facility was built near Carlsbad, New Mexico, and it began full-scale operations on March 26, 1999. This is a facility paid for entirely by the U.S. government to store radioactive waste left over or produced from government nuclear activities. Almost all of this material is from nuclear weapons production or experimentation, and it does not include any commercial power production waste.

The radioactive material is sequestered in rooms 2,150 feet (655 m) underground, in the middle of a salt formation 3,000 feet (1,000 m) thick. This salt, in the Salado and Castile formations, has been stable for more than 250 million years and is not likely to change in the future. Waste

A 2010 photo of radioactive waste being stored in the WIPP facility in Carlsbad, New Mexico. Waste from the nuclear weapons program is inserted into holes drilled in the walls of an ancient salt formation one-half mile underground.   (© *Jim West/The Image Works*)

disposal operations are expected to continue until 2070, at which time the vertical access tunnel will be sealed permanently. By 2006, the facility had buried 5,000 truckloads of waste from nuclear weapons plants as they were being shut down and dismantled.

Since the beginning of the commercial nuclear power industry in 1957, the U.S. government has imposed a tax of 0.1 cent per kilowatt-hour of power produced by nuclear means. There are 65 nuclear power plants in the United States, with a total of 104 reactors. The typical power plant makes electrical energy at a rate of 1 billion watts. Running all day and all night at full power, a nuclear plant makes 24 million kilowatt-hours in a day, and for that it owes $24,000 in tax. The revenue collected from this special tax, accruing for more than 50 years, is to be used by the government to collect and dispose of all the spent fuel from the power plants. The government collects from $300 to $500 million per year from the nuclear power industry, specifically for waste repository. The U.S. Navy needs a long-term storage facility for reactor waste from submarines and aircraft carriers and has chosen to participate in the funding. The navy pays about 27 percent of the total waste disposal cost, taken from its taxpayer-funded budget.

It was a fine idea for the government to take responsibility for the waste and to be paid for it, but progress has been slow. So far, more than 150 million pounds (70 million kg) of spent fuel have piled up at commercial nuclear power plants, waiting for the trucks to come and haul it away. The tax is collected regularly. After studying the problem for 20 years, in 1978, the DOE focused its investigation into the ideal burial spot for fission products to a ridgeline of volcanic material in south-central Nevada. It is in an uninhabitable desert owned by the federal government, adjacent to the Nevada nuclear weapons test area in Nye County. It is Yucca Mountain.

In 1982, the U.S. Congress crafted and passed a new law called the Nuclear Waste Policy Act, making the burial of nuclear fission waste an official policy. By 1984, the DOE had selected 10 possible locations for the repository, and in 1985, the choice was narrowed down to three candidates: Hanford, Washington; Deaf Smith County, Texas; and Yucca Mountain, Nevada. The Nuclear Waste Policy Act was modified in 1987 to consider only Yucca Mountain, and test drilling and experimentation were concentrated in this one spot.

An aerial view of Yucca Mountain, Nevada   *(U.S. Department of Energy/Photo Researchers, Inc.)*

Initial testing and extensive geological investigation found the location acceptable, and the DOE announced that it would begin accepting spent fuel at the Yucca Mountain Repository by January 31, 1998. The date was later than the commercial power plants would have preferred, but at least the site would be thoroughly investigated and carefully constructed. More room for used fuel to be stored at the plants would have to be set aside while the Yucca Mountain burial site was prepared.

A special tunnel boring machine, 400 feet (125 m) long, was built at a cost of $13 million, and the main tunnel was dug out of the mountain. The excavation is U-shaped, five miles (8 km) long and 25 feet (8 m) wide. Cross tunnels are to be dug on an as-needed basis. The year 1998 passed without completion of the facility, and after that the U.S. government was out of compliance with its contract with the nuclear industry and technically owed the money collected as taxes. On July

18, 2006, the DOE proposed March 31, 2007, as a new opening date. By this time, 900 people were employed on the project, and in 2007 DOE announced plans to double the size of the Yucca Mountain Repository to a capacity of 300 million pounds (135 million kg) in anticipation of a greater need for waste storage. As of 2008, $9 billion had been spent on the project, and Yucca Mountain was the most studied piece of geology in the world. The total cost estimate for completion and long-term management was $90 billion. The opening date was pushed to September 2020.

Using Yucca Mountain as a burial site for fission products has been a controversial topic in Nevada for several years, but in 2006 opposition to the project became significant when Nevada senator Harry Reid (1939–) became majority leader. Blocks on project funding and an NRC operating license threatened to stall the project beyond the point where it would be usable as a burial site, and power plants would have no choice but to store spent fuel for another decade or two while the problems are resolved. It may be necessary to simply cover the tunnel and start over in some other state. By any means, the long-term storage of spent fuel must be accom-

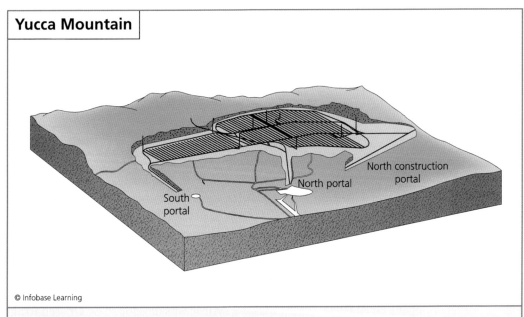

Yucca Mountain

North construction portal

North portal

South portal

© Infobase Learning

An oblique view of Yucca Mountain with the earth cut away, showing the intricate series of main tunnels and cross tunnels in which nuclear reactor waste is to be stored.

In this April 13, 2006, photo, Pete Vavricka conducts an underground train from the entrance of Yucca Mountain in Nevada. *(AP Images)*

plished if nuclear fission is to continue as a power source. There are other ways to make storage possible, but none are as well developed as simply burying it underground.

SYNROC: MAKING SYNTHETIC RADIOACTIVE ROCKS

An often-stated reason for opposition to buried nuclear waste is the possibility of it getting into the drinking water supply. If a storage canister breaks and a tunnel floods, it could be possible for soluble fission products to dissolve in the underground water. A way to make this impossible is to embed the waste material in synthetic rock, called *synroc*. Made partly of a hard, rocklike material, radioactive waste would be not be likely to get in the water supply. It would not readily dissolve.

A practical process for making synroc was developed by Dr. Ted Ringwood (1930–93) and his team at the Australian National University in Canberra in 1978. Three titanate minerals comprise Ringwood's synroc:

hollandite, zirconolite, and perovskite, with a small amount of titanium dioxide and a metal alloy. These chemicals are combined in a slurry of water and a portion of high-level nuclear waste is mixed in. The mixture is dried and calcined at 1,380°F (750°C), producing a powder. The powder is compressed at 1,102–2,192°F (1,150–1,200°C), and it becomes a hard, dense, black rock. The radioactive waste is locked into the crystalline structure of the synroc, as if it were a naturally occurring mineral. The synroc can withstand heavy radiation flux within its structure without breaking down, cracking, or reverting to powder form, and these properties make it an ideal way to immobilize fission products while they decay to a nonradioactive form. It resists leaching into groundwater, but it must be buried. Although the material is locked into the rock, it remains dangerously radioactive.

Synroc is being used to process nuclear waste by the DOE at the Idaho National Laboratory and is part of a demonstration contract at the British National Fuel Facility at Sellafield, England, to immobilize plutonium-contaminated material.

VITRIFICATION: MAKING RADIOACTIVE GLASS

An alternative to making synroc is to embed the fission products in glass. While not quite as stable as synroc, it is easier to do and uses commonly available materials. The process, called *vitrification,* has been used successfully to convert nuclear waste in any form into cylinders of colored glass.

Vitrification involves solidifying silicon dioxide, or powdered quartz, without allowing it to revert to its crystalline form. With no crystallization, the silicon dioxide has no preferred fracture planes, and it has the properties of a fluid that cannot move. The silicon dioxide, in the form of crystalline sand, is melted, and sodium carbonate and calcium oxide are added to prevent recrystallization as it slowly cools. If radioactive waste is mixed in, then it becomes part of the glass and is immobilized in the matrix of silicon dioxide molecules. When cooled, it is a hard, black material, resembling obsidian. It is insoluble in water, the waste cannot be washed or blown away, and it is expected to be stable for hundreds of years. Like synroc, the glass is radioactive, and it must be buried or otherwise removed from biological contact.

Experiments with vitrification have shown it to be a relatively simple, cost-effective way to immobilize waste, but there may be a weakness. Radiation in the silicon dioxide matrix, particularly alpha radiation, tends to break down the structure and turn the glass into powder. Further development may be necessary to make this material suitable for long-term storage of fission products.

ION EXCHANGE: CONCENTRATING RADIOACTIVITY INTO A SMALL VOLUME

Spent fuel from a power reactor is usually less than 5 percent radioactive fission products. The rest of it is unburned uranium-235, non-fissile uranium-238, and plutonium-239. These transuranic elements can be reused in the power production process, and as the price of reactor fuel rises as a world commodity, it will become impractical to discard these valuable components of spent fuel. Another factor is that the volume of the spent fuel falls dramatically if the transuranics are removed, making it easier and less space-consuming to permanently bury the true waste products.

A proven method of concentrating these materials is called ion exchange, in which charged particles are traded between two electrolyte solutions. By then separating the solutions, selected elements can be extracted from complex mixtures. There are many proven ion-exchange methods being used worldwide for spent fuel processing, such as UREX, TRUEX, DIAMEX, SANEX, and UNEX, but all are variations of the original process, the plutonium-uranium extraction, or *PUREX,* invented by American scientists in the course of atomic bomb production during World War II. It was first used to extract plutonium-239 from spent reactor fuel at the Hanford Plant in Richland, Washington.

In the PUREX process, spent reactor fuel is dissolved in nitric acid. Fine, insoluble solids are then filtered out of the solution. A mixture of tributyl phosphate and kerosene is mixed with the acid solution, and the uranium and plutonium remain liquid and dissolved in the kerosene while the fission products fall out of solution. The kerosene containing the transuranic elements is then drawn off the acid and mixed with water. Plutonium is separated out of the kerosene, which tends to float atop the water, by treating it with ferrous sulphamate. Plutonium then dissolves in the water, and the uranium can be back-extracted separately from the kerosene using more nitric acid.

This extraction must be repeated, because the extreme radioactivity in the fission products causes a degradation of the process. Under radiation,

THE CLOCK OF THE LONG NOW: TECHNOLOGY TO LAST FOR 10,000 YEARS

A problem with long-term fission product storage facilities is that the length of time required for safe storage can run into tens of thousands of years. There is no guarantee that the United States or anything resembling it will still be here 10,000 years from now, but our nuclear waste will still be buried in the ground. There will be remnants of radioactivity that nobody should be exposed to without knowing what they are and how dangerous they may be. Someone in the far future could discover a waste burial site and think that it is an important treasure cache or a time capsule left by an ancient civilization. There could be a temptation to dig it up.

Anticipating this, a team of linguists, scientists, science fiction writers, anthropologists, and futurists are trying to design a warning system for the Waste Isolation Pilot Plant in New Mexico. This is a difficult assignment. People in 10,000 years will not necessarily read language or know what a danger symbol looks like.

A similar challenge is being met by the Long Now Foundation, and their extensive research may point to methods that may be used for a warning system. The term *long now* comes from cosmology, which divides the march of time into two periods. The first period extended from the instant of creation, or the beginning of the Big Bang, up to a few minutes into the expansion of the universe. Time, space, and the laws of physics were not fixed and stable in this period, particularly in the first 10^{-37} seconds, and things were in extreme flux. After this first, massive expansion period, the universe began to settle down. Space became mostly vacuum, and the fundamental particles of matter were established. The universe has been this way basically for the past 13.75 billion years. This second period is called the long now. When considering very lengthy passages of time, the long now is an appropriate descriptor.

An important project of the Long Now Foundation, conceived by Danny Hillis (1956–), is to build a clock that will keep time and count off the years, chiming once for each millennium. This timepiece is supposed to last for 20,000 years, which is several times longer than any human-made object has endured through history. It is to be impressive and well engineered, drawing people to it just to see it tick, and causing them to admire it and think about the ancients that built it. It must remain accurate for thousands of years, it must not be made of valuable materials, for fear

(continues)

(continued) _____

of looting, and its power source must be constant over its lifetime. It must be driven by a falling weight, and admirers will have to wind it. It must be easy to maintain and repair. These are tall specifications, and much thought has gone into its design. The foundation has bought a building site on top of Mount Washington near Ely, Nevada, near a number of 5,000-year-old bristlecone pines. The Clock of the Long Now will be built mainly underground, accessible only by foot traffic. The first working prototype is on display at the Science Museum in London.

The warning system at WIPP must have similar longevity, and the problem is to prevent it from being an enticement. It will include an outer perimeter of 32 granite pillars, each 25 feet (7.6 m) tall. Inside the perimeter will be a circular earthen wall and within the wall will be another 16 granite pillars. At the center will be a granite, roofless room, with warnings etched into the walls, written in seven languages including Apache.

The final plan is to be submitted to the DOE by 2028.

the tributyl phosphate and kerosene mixture produces dibutyl hydrogen phosphate, and this causes the transuranic elements to be contaminated with fission products. With each pass through the process, the radiation level is reduced and the extraction purity improves.

Care must be exercised in applying ion-exchange processes. It is possible to create a nonscheduled critical mass in a plutonium or uranium solution, and people have been killed in such accidents. The fission products collected in the nitric acid solution cannot be treated lightly and must be fully contained under heavy shielding. In the early days of weapons production, environmental safety was not a first concern, and it is estimated that 685 kilocuries (25.3 PBq) of radioactive iodine-131 were released into the atmosphere and the Columbia River at the Hanford Plant from 1944 through 1947. Cleanup is still ongoing, at an estimated cost of $2 billion per year.

There are presently no ion-exchange spent fuel reprocessing plants for commercial reactors in the United States, but there are one in China, three in France, two in the United Kingdom, three in India, two in Japan, one in Russia, and one in Pakistan.

TRANSMUTATION OF ACTINIDES

Most of the troublesome results of making power using fission are the fragments of the uranium nucleus, made radioactive by their overabundance of neutrons. These atoms eventually revert to stable forms, usually in more than one decay increment, emitting radiation with each step. There is another class of radioactive waste from fission, the transuranic elements resulting from the nonfissioning activation of uranium. Occasionally, a uranium-235 nucleus will not fission upon capture of a neutron, and uranium-238 nuclei never fission. These nuclei simply capture the neutrons, and the beta-minus decay products of the results become the nuclei of still heavier elements. These elements are in the class *actinides* on the periodic table of the elements. Examples of actinides made in power reactor fuel are neptunium, plutonium, californium, americium, and curium. All are radioactive, and, because they are more than twice as heavy as any product of fission, their decay chains down to stable elements are unusually long. They are, in fact, the last components of nuclear fission waste to reach stability. If the actinides could be eliminated from the waste, then the period of dangerous radioactivity from reactor waste could be shortened. Most fission products have half-lives of 90 years or less, while actinides have half-lives in the 100,000 to 200,000 year range.

It is possible to transmute actinides, or cause them to capture neutrons in an ongoing fission process and either fission, contributing to the power process, or activate into rapidly decaying versions. In the process of transmuting the actinides, the plutonium component is fissioned, resulting in increased power output and half-sized plutonium fission products. This transmutation requires fast, high-energy neutrons, so it does not occur to a significant extent in thermal reactors used in the power industry. Fast, sodium-cooled breeder reactors are designed specifically to not moderate the neutron speeds, and this type of reactor is necessary for the transmutation effect. The Integral Fast Reactor project, started in 1983, was intended to specifically demonstrate the transmutation of actinides in reactor fuel, but the program was halted by the U.S. Congress in 1993. Sodium-cooled fast reactors having transmutation capability are among the Generation IV reactors currently being designed.

The actinide reduction issue is considered important in the European Union, and at least two major research groups, *ACTINET-I3* and *ACSEPT,* are presently investigating industrial scale transmutation. ACTINET-I3 has three research centers in France, three in Germany, one in Switzerland, one in Sweden, and one in the United Kingdom working on the problem of actinide transmutation. Other European Union participants are the Czech Republic, Cyprus, Poland, and Russia. This impressive collaborative effort is using these well-equipped research facilities to pool resources and perfect an industrial system that will design and build everything from the special fast reactors to the waste reprocessing systems.

Related to ACTINET-I3 is a research and development project named ACSEPT, funded by the *Seventh Framework Programme (FP7)* of the European Commission. This project, focused on the actinide chemistry of spent reactor fuel, began on March 1, 2008, and is expected to run until February 29, 2012. The research consortium consists of 12 European countries plus Japan and Australia. ACSEPT is working to produce the design of an advanced fuel processing plant that will separate out actinide components, such as plutonium, that are useful as thermal reactor fuel, and cycle the unusable portions back into fast reactors for transmutation. Anticipating that an appropriate Generation IV reactor will be adopted for actinide reduction, this scheme will minimize the amount of spent fuel that will have to be buried or vitrified. At *SCK·CEN,* a nuclear research center in Mol, Belgium, a fast-spectrum research reactor named *MYRRHA* is being designed to test methods of actinide transmutation. It is expected to be operating at full power, 100 megawatts, by 2023.

Presently, all the power reactors in Europe produce about 2,800 tons (2,500 metric tons) of spent fuel per year, containing 28 tons (25 metric tons) of plutonium and 3.9 tons (3.5 metric tons) of other actinides.

MAKING CONSTRUCTIVE USE OF NUCLEAR WASTE PRODUCTS

Some fission products have wide application in industry, and they should be extracted from fission waste in an organized fuel-reprocessing program. The uses for radioactive cobalt-60 and cesium-137, which are abundant in reactor waste, include medical sterilization, food sterilization, cancer treatments, and industrial radiography. The use of these nuclides is tightly regulated, as they are intense gamma-ray emitters and constant vigilance and shielding are necessary, but their advantageous use is undeniable. Gamma ray sources are used to spot faulty welds in field construction of pipe systems, off-shore drilling platforms, power plants, and industrial systems of all descriptions, significantly reducing the chances of accidental pipe ruptures.

Technitium-99m, the metastable form of technetium-99, is among the most widely used radiopharmaceuticals in the world. It is a fast-decaying gamma source, used in approximately 200 million diagnostic procedures every year, and 85 percent of all medical imaging uses this isotope. It is used for bone scans, myocardial perfusion imaging, cardiac ventriculography, functional brain imaging, immunoscintigraphy, blood pool labeling, and spleen scanning. It is derived from the fission product molybdenum-99. Other important radiopharmaceuticals are yttrium-90, used to treat arthritic conditions and prostate cancer, and iodine-131, used to treat thyroid cancer. These isotopes are available only in spent reactor fuel.

At this time, most of the radioactive medical isotopes used in North America come from a single source, the NRU reactor at Chalk River, Canada. This reactor has been running almost continuously since November 3, 1957. When it is finally shut down, there is nothing standing by to replace it. Most of Europe's supply comes from the Petten reactor in the Netherlands, and this valuable resource has been running since 1960.

There is much work left to be done, particularly in the area of nuclear power waste disposal. Research is ongoing at a greater pace than ever before in fission product sequestering, as well as advanced reactor design and some very far-reaching applications of nuclear reactions. The industrial world is preparing for a future that will inevitably include power production by nuclear means.

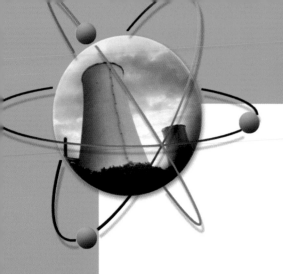

Conclusions

Europe has always seemed a step ahead of North America when it comes to energy crises. In those days so long ago when gasoline cost 32 cents a gallon in the United States, it was already $4 a gallon in the Eastern Hemisphere. Coal production and the severe pollution caused by burning it hit Europe early, and by the end of World War II France and England were looking for alternate ways to produce electricity and heat homes. England responded by building the world's first high-capacity commercial nuclear power plant, Calder Hall, beginning construction in 1953. Although it was clear from the beginning that nuclear power would be the more expensive way to derive electricity, at least it did not involve mining and burning coal. France was not far behind, eventually becoming the country having the highest percentage of dependence on nuclear fission for power. About 80 percent of the electricity in France comes from nuclear fission. The European Union now has plans into the far future, expanding the nuclear power net as well as wind, geothermal, and solar power. Anything but coal is under consideration.

A pause in this progress may result from the most recent earthquake and tsunami in Japan. It caused the destruction of an entire six-unit power plant on the east coast, but this disaster will only lead to more design rigidity, focus on seismic shock, and refined emergency measures. The march forward, even in Japan, will continue at pace.

These long-term plans include a commitment to fusion power. The largest, most powerful fusion reactor ever designed, the ITER, is now being assembled in Cadarache in the south of France. Japan, China, South Korea, and Russia, as well as the entire European Union, are involved in this project. It is designed to generate 500 megawatts of power and is expected to begin tests in 2018. Hedging bets against fusion technologies, there is also a completely different large-scale fusion development being built in Bordeaux, named Laser Mégajoule. It should be ready for testing in 2012.

The European Union, China, Japan, and India are all committed to the highly advanced Generation IV reactor designs, and France, Japan, and Russia have already developed and run high-output sodium-cooled fast breeder reactors in anticipation of a not-distant time when uranium will become a scarce and expensive commodity. India has already developed a thorium fuel cycle and is designing reactors to burn it as fuel. A thorium reactor is under construction, and China and Japan are close behind. Given a thorium fission technology, India, which now has the world's second highest population and a rapidly developing economy, can become independent of external power sources. China and Japan are watching this development closely and are working on their own thorium reactor designs as part of Generation IV.

Russia is also working to build nuclear reactors for future economic growth. Their new line of VVER reactors includes a core catcher built into the base of each reactor containment structure, designed as a last-ditch effort to keep a melted core from becoming a bigger problem in a major reactor cooling system breakdown. The steaming, bubbling blob of fuel, reactor structures, and molten control rods are collected at the bottom in a neatly arranged cup, ready for planned disposal. Many nuclear engineers in the United States looked askance at this design feature. It seemed to indicate that the reactor was likely to fail, and they were designing it expecting it to eventually melt down. With the meltdown of three reactors designed in the United States at Fukushima I in Japan, this concept no longer seems like a bad idea. Russian nuclear engineers, Chinese nuclear engineers, and engineers in both the Eastern and Western Hemisphere have learned to pick and choose the best ideas from everyone's reactor designs, as well as study closely each mistake, disaster, and catastrophe in order to build a better, safer reactor. The VVER core catcher may show up in future Generation IV designs.

The United States has never been alone in its pursuit of commercial nuclear power, but for decades it was at the head of the development curve, pioneering concepts, designs, and experimental facilities. This is no longer the case. This country, despite the fact that it has the most reactors, 104, now lags far behind its economic competition in Europe and Asia. Every power reactor in the United States is a Generation II facility, built with technology that is now 40 years old. In a world where cell phone technology is turned around every six months, our nuclear technology is antique. These power plants will come to the end of their useful lives, one at a time, over the next two decades, and now is the time to begin the replacement program, just to maintain our 20 percent nuclear power fraction and not to replace the majority power source, which is burning coal.

An expansion of the nuclear power network has started in the United States, with a ground breaking at Plant Vogtle in Georgia to build two Generation III+ reactors. They represent the first nuclear construction start in 30 years, and the reactors are Westinghouse AP1000s. The Westinghouse Electric Company is no longer the homegrown firm it used to be. It is now owned by Toshiba of Japan, and the industrial infrastructure of the United States can no longer forge a steel object as big as a reactor vessel. The parts for Plant Vogtle reactors will be manufactured in Japan.

All is not lost. The age of the mammoth, 1 billion watt power reactors may, in fact, be a passing step in the evolution of nuclear power. There is a new push for small modular reactors. Instead of having a huge power plant sending electricity to distant cities, it may make more sense to have a carpet of tiny reactors, closer to the consumer points. When a big reactor fails, melts down, and blows up the building, it is a proportionately big disaster. When a tiny reactor fails, the result is a tiny disaster, easier to clean up and safer for the surrounding population. Large reactor plants have to be built in place, with a great deal of pipe-fitting, wiring, and concrete pouring. It can take a decade of construction effort. A small reactor can be built in a factory, on an assembly line, as if it were an automobile. It can be transported to the plant site on a truck or a railcar, lowered into a prebuilt silo, connected to the power gird, and turned on. The United States is at the forefront of developing this technology, and it could turn out to be the next big thing in power generation.

In all, there is a strong future for nuclear power. There will be triumphs, breakthroughs, and incremental steps as more is learned and the art continues to progress. There will be accidents, disasters, and acts of God. Lessons will be learned, and the efforts to build absolutely safe power plants will continue, even if the goal is ultimately unachievable. There will be an ever-growing need for science, engineering, knowledge, and imagination in this fascinating corner of the technology of civilization.

Chronology

1867   Babcock & Wilcox begins business by patenting the nonexplosive steam boiler in Providence, Rhode Island.

1870   Mitsubishi Heavy Industries, Ltd., of Tokyo, Japan, is established as a shipbuilder in Nagasaki.

1886   George Westinghouse (1846–1914) founds the Westinghouse Electric Company in Pittsburgh, Pennsylvania.

1890   Thomas A. Edison (1847–1931) starts the General Electric Company in Schenectady, New York.

1910   Hitachi Ltd. of Tokyo, Japan, is established as an electrical repair shop.

1912   Combustion Engineering is formed by the merger of two makers of fuel-burning equipment in New York City.

1939   Tokyo Shibaura Denki is formed by the merger of two electrical machinery makers in Japan. It is nicknamed "Toshiba."

1957   **December 23** Nuclear power becomes a commercial venture when the Shippingport Atomic Power Station in Pennsylvania begins producing electricity for sale to consumers.

1958   **May 3** The TRIGA Mark 1, a reactor safe enough to be run by graduate students, is licensed for operation by the Atomic Energy Commission in San Diego, California.

1960   Construction begins on the world's first pebble-bed reactor, the AVR, at the Jülich Research Centre in West Germany.

1961   **May 14** The Tory-IIA, the world's first nuclear-powered ram jet, is successfully tested at Site 401 in Jackass Flats, Nevada.

1965   The Molten Salt Reactor Experiment is first run at full power in Oak Ridge, Tennessee.

         April The Poodle radioisotope rocket is run for 65 hours at TRW.

1968   **June** The Phoebus 2A nuclear rocket engine develops 4,082 mega-watts of heat for 12.5 minutes, making it the most powerful nuclear reactor ever built and tested.

October The Molten Salt Reactor Experiment is the first reactor to run on pure uranium-233.

1970   **March 5** The Nuclear Non-Proliferation Treaty, signed by 40 nations, becomes effective.

1983   **February 1** The first Generation III CANDU 6 power plant begins service in Point Lepreau, Canada.

1985   The first of six VVER-1000 Generation III reactors at the Zaporizh-zhia Power Station, Ukraine, begins commercial operation.

1987   **September 29** GE submits the first section of the Standard Design Certification Application for the ABWR to the NRC.

1988   The ABB Group of Zürich, Switzerland, is formed by the merger of ASEA (Sweden) and Brown, Boveri & Cie (Switzerland).

1989   **November 18** The Babcock & Wilcox Nuclear Power Division joins with Framatome SA in France.

1994   The DOE's Integral Fast Reactor project is cancelled three years before completion.

1997   **May 12** The NRC issues a final rule certifying the GE ABWR design.

May 20 The ABB System 80+ reactor is given design certification by the NRC.

1998   British Nuclear Fuels Limited buys the Westinghouse Electric Company.

2000   **January** The U.S. Department of Energy convenes a meeting of nine countries to begin the development of the Generation IV nuclear power systems.

2001   **July 16** A representative of the United States signs the Generation IV International Forum Charter.

September 3 AREVA of Courbevoie, France, is formed by the merger of the French companies Famatome SA, Cogema, and Technicatome.

2002   **February 14** The Nuclear Power 2010 Program is unveiled by the U.S. secretary of energy Spencer Abraham (1952–) as an incentive for developing new reactor technology.

2003   **July 24** The last Generation III CANDU 6 reactor begins service at the Qinshan Power Plant, China.

2004   **December 14** The city council of Galena, Alaska, votes to allow Toshiba to install a 4S modular reactor for testing its use in a remote area.

2005   **February 17** A construction permit is issued to TVO, Finland, for the Olikiluoto Unit 3 EPR reactor, the first Generation III+ reactor to be built.

        August 24 GE-Hitachi submits the application for Standard Design Certification for the ESBWR to the NRC.

2006   **January 27** The Westinghouse AP1000 Generation III+ reactor is issued a final rule certifying the design.

        June 19 NRG Energy of Texas files a letter of intent with the Nuclear Regulatory Commission to build two Generation III ABWRs at the South Texas Project site.

        October 17 Toshiba completes its acquisition of Westinghouse from BNFL.

2007   **June 4** General Electric and Hitachi form GE-Hitachi and the Global Nuclear Alliance.

        December 7 Work begins on the AREVA Generation III+ EPR at the Flamanville Nuclear Power Plant, France.

        December 11 AREVA files an application with the NRC for a Standard Design Certification for the EPR.

2008   **February 27** Grand Gulf applies to the NRC for a Construction and Operator's License for a GE-Hitachi ESBWR.

        April 9 The Georgia Power Company signs a contract with Westinghouse to deliver two AP1000 Generation III+ reactors to Plant Vogtle.

        October 25 Construction begins on the Leningrad Nuclear Power Plant II in Russia. The plant will have two Generation III+ VVER-1200 reactors.

2009	**July** B&W begins preapplication design certification interaction with the NRC to discuss the mPower modular reactor.
	December 9 KOPEC of South Korea sells four APR-1400 reactors to the United Arab Emirates.
2010	**March 23** Toshiba files an intent to ask for design approval with the NRC for its 4S modular reactor.
2011	**January 24** A French consortium announces a two-year feasibility study for a submerged modular nuclear power plant called Flexblue.
	March 11 A major earthquake and tsunami in Japan proves that the Generation II reactors are obsolete. The failure of the diesel generators at Fukushima I result in the destruction of the entire plant.
	July 22 The Nuclear Regulatory Commission issues a report documenting decades of scientific and technical work on the dormant waste repository in Nevada at Yucca Mountain. No conclusions or recommendations are made.
2015	**July** The nuclear-powered spacecraft *New Horizons* encounters Pluto and its three moons.
2016	The first ACR-1000 Generation III+ reactor is planned to be in operation somewhere in Canada.
	March In Japan the Tsuruga-3 Generation III advanced pressurized water reactor is brought to criticality for the first time.
2018	The first NuScale modular reactor power plant becomes operational.
	The ITER fusion reactor in France begins operation.
2020	A test version of the traveling wave reactor, the TP-1 built by TerraPower, begins operation.
	The Yucca Mountain nuclear waste repository may open for business.
2023	The MYRRHA fast-spectrum test reactor at SCK7CEN in Belgium begins full-power testing of actinide transmutation methods.
2028	The final plan for the WIPP warning system, designed to keep future generations from digging up the buried radioactive waste, is submitted to the DOE.
2030	Generation IV reactors take over as the dominant designs of nuclear power systems.

2049   **February 9** The operating license for the Unit 2 Generation II West-inghouse reactor at Plant Vogtle, Georgia, runs out.

2070   The WIPP waste disposal facility in Carlsbad, New Mexico, stops accepting radioactive material and is sealed.

2073   The AREVA Generation III+ EPR at Flamanville, France, is expected to shut down after 60 years of producing electrical power.

2170   Aboveground radioactivity monitoring at the WIPP ceases, as it is no longer necessary.

Glossary

4S   the super safe, small and simple reactor being developed by Toshiba; its first use will be to power the small village of Galena, Alaska.

ABB Group   one of the largest engineering conglomerates in the world, jointly owned in Switzerland and Sweden, and headquartered in Zürich, Switzerland.

ACSEPT   a large research project, funded by the European Union, to design a pilot plant for separating minor actinides from fissile actinide isotopes

ACTINET-I3   a European research consortium, presently developing an actinide transmutation fuel cycle

actinides   the subseries of elements on the periodic table of the elements beginning with thorium, element number 90, including uranium and plutonium, and ending with lawrencium, element number 103

advanced boiling water reactor (ABWR)   a Generation III design by General Electric

advanced heavy water reactor (AHWR)   developed at the Bhabha Atomic Research Centre (BARC) in Mumbai, India, it uses heavy water as the neutron moderator and pressurized light water as the primary coolant.

advanced power reactor 1400 APR1400   developed by Korea Hydro & Nuclear Power Company (KHNP) upgraded to produce 1,400 megawatts of power.

advanced pressurized water reactor (APWR)   a Generation III design built by Mitsubishi

Aircraft Nuclear Propulsion Program (ANP)   initiated by the air force in the 1950s to develop a nuclear powered strategic bomber

aneutronic   a nuclear reaction that produces no free neutrons

AP1000   the Westinghouse advanced pressurized water reactor with passive emergency core cooling. The AP1000 makes 1000 megawatts of power.

Arbeitsgemeinschaft Versuchsreaktor (AVR)   first experimental high-temperature pebble-bed reactor, built in Westphalia, Germany, in 1960

AREVA   a multinational construction conglomerate, based in Tour Areva, France. AREVA owns the Babcock & Wilcox nuclear division.

Army Nuclear Power Program (ANPP)   a project started in 1954 to supply the army with portable and stationary small nuclear power sources

Atomic Energy of Canada Limited (AECL)   a Canadian federal Crown corporation with the responsibility of managing Canada's national nuclear energy research and development program. AECL developed the CANDU reactor in the 1950s.

Bhabha Atomic Research Centre (BARC)   based in Mumbai, India, where the thorium fuel cycle is being designed here

CANDU 6   an advanced Generation III version of the CANDU reactor, being built by AECL

CNO   the carbon-nitrogen-oxygen fusion reaction that occurs in heavy stars. It is considered as a fusion mode in interstellar space vehicles because of its large reaction cross section.

containment   one of a short series of barriers to escaping radiation built into power reactors. The first containment is the reactor vessel, the second is a specially designed structure around the reactor vessel, and the third is the reactor building.

economic simplified boiling water reactor (ESBWR)   designed by GE-Hitachi as a Generation III+ power plant

European Atomic Energy Community (EURATOM)   composed of members of the European Union, but distinct from it. Its purpose is to develop nuclear energy and sell the surplus electricity to non-EU countries.

European Pressurized Reactor (EPR)   known internationally as the evolutionary power reactor a Generation III+ power plant designed by AREVA

Seventh Framework Programme (FP7)   the European Union's chief portal for project funding for research and technical development in the period from 2007 to 2013. The main scientific project in FP7 is the ITER fusion reactor.

gas-cooled fast reactor (GFR)   Generation IV concept that drives a Brayton cycle turbine directly from the reactor coolant

Generation IV International Forum (GIF)   convened in 2000 by the DOE to kick off the advanced reactor design effort

high temperature gas reactor (HTGR)   a design first proposed at the Oak Ridge National Laboratory in 1947. This was the first use of the pebble-bed concept, in which fuel is in loose ceramic spheres instead of in a rigid, geometric matrix.

Hitachi   a Japanese manufacturing company headquartered in Tokyo, Japan. Hitachi has partnered with General Electric, for designing, building, and selling new BWR reactors and power plants.

Hyperion   a small, inexpensive modular reactor that uses uranium hydride fuel. The uranium hydride compound imparts a negative thermal coefficient of reactivity that will shut the reactor down if the temperature rises too high.

Korean standardized nuclear plant (KSNP)   a type of PWR built to a standard design in South Korea. Six power plants of this design, built almost entirely with homegrown parts, are operating in South Korea.

lead-cooled fast reactor (LFR)   Generation IV design that produces hot gas in the final cooling stage. The hot gas is to be used to power a Brayton-cycle turbine, without the use of steam.

magnetic target fusion (MTF)   a nuclear propulsion technique used in by NASA in the *HOPE* spacecraft design. In this concept, particle accelerators are used to heat light isotopes to fusion temperatures.

Mitsubishi Heavy Industries, Ltd.   a heavy industrial manufacturing company headquartered in Tokyo, Japan, now building APWRs for electrical power production in Japan.

molten salt reactor (MSR)   a Generation IV design that will use thorium fuel dissolved in salt made liquid by the extremely high running temperature

Molten Salt Reactor Experiment (MSRE)   the first large-scale use of molten salt fuel in a power reactor prototype, at the Oak Ridge National Laboratory in 1964. It was the first use of a Haynes alloy named Hastelloy-N.

mPower   a compact, modular reactor designed by B&W. The mPower is considered a Generation III++ design. It is more advanced than a Generation III+, but the design is complete and tested, which means it is not a Generation IV.

multi-application small light water reactor (MASLWR)   developed by Oregon State University and the Idaho National Laboratory starting in 2000. It is the basis for the NuScale modular reactor.

MYRRHA   a fast-spectrum research reactor, being designed to test methods of actinide transmutation at SCK·CEN in Mol, Belgium. Maximum power will be 100 megawatts.

Nuclear Regulatory Commission (NRC)   an independent agency of the U.S. government headquartered in Rockville, Maryland. Established in 1975, the commission oversees nuclear safety, licensing, and fuel management.

NuScale   a small, modular power reactor being developed by NuScale Power, Inc., in Covallis, Oregon. It will generate 40 megawatts of electricity using conventional, readily available reactor fuel. This feature separates it from other modular reactor designs, in that it will not require special, expensive fuel fabrication.

PUREX   an ion-exchange method used to extract plutonium and uranium from spent reactor fuel, as developed for the Hanford site in Richland, Washington, in 1947.

QUADRISO   an advanced form of fuel pellet, developed at the Argonne National Laboratory. A layer of europium oxide or erbium oxide is deposited between the fuel and the first carbon layer, acting as a burnable neutron poison. The use of burnable poison allows for more fuel to be loaded into the reactor than is needed for running at power without overloading the controls, and it extends the time between refuelings.

radioisotope thermoelectric generator (RTG)   a device that generates DC current using a radioactive heat source and a thermocouple. These devices have no moving parts and are extremely reliable. They are used in all deep-space exploratory missions.

reactor in-flight test (RIFT)   a hot-run of a spaceborne nuclear power reactor or engine in space

reaktor bolshoy moshchnosti kanalniy (RBMK)   A graphite-moderated nuclear power reactor; in English, translated as high-power channel-type reactor. This Russian-built reactor was the one involved in the Chernobyl disaster.

Rosatom   The Russian state manufacturer of nuclear fuel. About 17 percent of all the nuclear fuel used in the world is made and sold by Rosatom.

SCK·CEN   a combination Dutch-French abbreviation for Center for the Study of Nuclear Energy. The center is located in Mol, Belgium, employing approximately 600 people. It has four nuclear research reactors operating and one in construction.

supercritical water reactor (SCWR)   a Generation IV design that uses one simple cooling loop leading directly from the top of the reactor vessel to the turbogenerator. It is different from the boiling water reactor in that the water does not boil, but is kept at a supercritical condition in which water and steam are indistinguishable.

sulfur-iodine cycle (S-I cycle)   an industrial process for manufacturing hydrogen gas from water, using only heat as an input energy. Iodine and sulfur dioxide are recycled continuously in the process, as intermediate chemical-reaction steps. Hydrogen production using the S-I cycle is seen as a application of the VHTR.

sodium-cooled fast reactor (SFR)   a traditional breeder reactor concept using two liquid sodium coolant loops in series. The old sodium-cooled reactor has been refined into a Generation IV design.

South Texas Project Electric Generating Station (STPEGS)   a nuclear power plant near Bay City, Texas. The plant, using two Westinghouse PWRs, began operation in 1988 and was the first nuclear power installation in Texas.

Space Nuclear Propulsion Office (SNPO)   opened jointly by NASA and the AEC to manage nuclear spaceflight development. Objectives and nuclear engine specifications were drawn up for test operations at the Nevada nuclear rocket test site at Jackass Flats.

synroc   a synthetic rocklike material, made by heat and compression applied to three titanate minerals. Synroc, which is mechanically stable and insoluble, can be used to immobilize spent nuclear fuel for possible disposal.

traveling wave reactor (TWR)   the active fission region of the reactor core migrates as fuel is used up on one side of the core and becomes active on the other side of the core. The TWR is a breeder reactor that does not require fuel processing or scheduled loading.

tristructural-isotropic (TRISO)   a form of pelletized reactor fuel used in pebble bed reactors. A TRISO pellet is typically the size a tennis ball and made primarily of pyrolytic carbon, with the middle composed of uranium oxide, uranium carbide, or uranium carbonate.

unicouple   a silicon-germanium semiconductor device that generates a direct current as heat is applied to one end of the junction while the other end of the junction is cooled.

US-APWR   the Japanese APWR, built by Mitsubishi, modified to comply with standards set by the NRC.

very high temperature reactor (VHTR)   designed as a Generation IV concept to run at unusually high thermal loading, it will be useful for generating industrial steam for several applications, including the manufacture of hydrogen gas.

vitrification   a method of immobilizing nuclear waste in glass. Pulverized spent is fed into a glass-making operation, and the resulting cylindrical chunks of glass can be sequestered until the radioactivity has decayed away.

Waste Isolation Pilot Plant (WIPP)   a nuclear waste repository located underground near Carlsbad, New Mexico. WIPP is entirely government owned and operated and used exclusively for nuclear weapons construction and experimentation on radioactive waste.

water-water energetic reactor (VVER)   a Russian design for a commercial PWR. A version of the VVER is considered a Generation III+ design, and is available in the international market.

Further Resources

BOOKS

Brand, Stewart. *The Clock of the Long Now: Time and Responsibility.* New York: Basic Books, 1999. The formation of the idea for the clock of the long now, an ambitious project to make a piece of technology that will last for 10,000 years, by the author of the *Whole Earth Catalog.*

Dewar, James A. *To the End of the Solar System: The Story of the Nuclear Rocket.* Ontario, Candada: Apogee, 2007. An excellent and complete account of the race for nuclear power in outer space, covering both the technical and the political challenges of developing rockets for manned spaceflight to Mars and beyond.

Dyson, George. *Project Orion: The True Story of the Atomic Spaceship.* New York: Henry Holt, 2002. The complete story of the *Orion* nuclear spacecraft, as told by the son of Freeman Dyson, who worked extensively on the project for General Atomics, using newly declassified material.

Gantz, Kenneth F. *Nuclear Flight: The United States Air Force Programs for Atomic Jets, Missiles, and Rockets.* New York: Duell, Sloan, and Pearce, 1960. A thorough overview of the state of the Aircraft Nuclear Propulsion program in 1960, just before it was cancelled and after much progress had been made. Included are a summary of the project's organization structure, the reasoning behind it, principles of the developed technology, a progress report, and special human considerations for the use of atomic powered airplanes.

Hore-Lacy, Ian. *Nuclear Energy in the 21st Century.* London: World Nuclear University Press, 2006. This lavishly illustrated summary of nuclear power includes many useful tables and graphs and gives a concise overview of the present state of world nuclear power, its history, and the future of Generation III+ and Generation IV reactor technologies.

Penner, S. S. *Advanced Propulsion Techniques: Proceedings of a Technical Meeting Sponsored by the AGRAD Combustion and Propulsion Panel, Pasadena, California, August 24–26, 1960.* New York: Pergamon Press, 1961. Concepts of nuclear rocket and jet propulsion were surprisingly well advanced by 1961, and this book covers everything from the Project Rover nuclear engines to magnetofluid-dynamic propulsion. This book is graduate-level reading, but it will be interesting to students studying the history and the interrupted progress toward interplanetary and interstellar travel.

Shapiro, Fred C. *Radwaste*. New York: Random House, 1981. Although this book is 30 years old, it still rings true and topical, as there has not been significant advancement of the art since it was written. As Shapiro says in his dedication to his children, "This is a fine mess we've gotten you into."

WEB SITES

Further depth on many topics covered in this book has become available on the Internet. Look for cross-referencing links in a Web site. A double click on a highlighted word can take you to even further, giving details on interesting subtopics.

ABWR Overview is a very informative, 56-page look at the unique design features of the General Electric Generation III advanced boiling water reactors. It includes specific design parameters, a power plant site plan, and detailed diagrams of the reactor and its auxiliary systems. Available online. URL: http://www.ne.doe.gov/np2010/pdfs/ABWROverview.pdf. Accessed November 1, 2011.

AECL ACR-1000 technical details are available on a pdf file from Atomic Energy of Canada. The 52-page document includes 49 color diagrams and renderings of the plant. Available online. URL: http://www.aecl.ca/Assets/Publications/ACR1000-Tech-Summary.pdf. Accessed November 1, 2011.

AP1000, the new Westinghouse Generation III+ reactor, is described in an informative series of YouTube videos. Watch the first one addressed here and then follow up with additional videos indexed at the site. Available online. URL: http://www.youtube.com/watch?v=R6T_AUZZeiw. Accessed November 1, 2011.

Bussard Fusion ramjets may be far in the future, but work is underway at this time to perfect the fusion reactor that will be necessary. This Web site, called "Building the OpenSource Bussard Fusion Reactor" is very interesting. It includes a video showing the Bussard reactor started up and running, with a description of the equipment needed to run it. Available online. URL: http://www.kickstarter.com/projects/1992078142/building-the-open-source-bussard-fusion-reactor. Accessed November 1, 2011.

Energy from Thorium is a Web site that gives a complete picture of the concept of using thorium as a reactor fuel. Included are lists of thorium-related videos, links, blogs, recommended books, e-books, and archives. Available online. URL: http://energyfromthorium.com. Accessed November 1, 2011.

ESBWR Overview is an excellent 47-page presentation of the Generation III+ economic simplified boiling water reactor presented by General Electric. Included are diagrams showing all the advanced features of the ESBWR power plant, including a site plan, cutaway diagram, and the cooling system. The operation of the advanced passive cooling system is shown in several pages of diagrams and explanations. Available online. URL: http://www.ne.doe.gov/np2010/pdfs/esbwrOverview.pdf. Accessed November 1, 2011.

Generation IV Nuclear Energy Systems are well described in this 36-page presentation given by Brookhaven National Laboratory at the University of Tennessee on April 30, 2003. It includes descriptions of all the Generation IV reactor concepts, from the very high temperature reactor to the supercritical water reactor, with high-quality color graphics. Available online. URL: http://alpha.chem.umb.edu/chemistry/ch471/evans%20files/ UTenn%20GenIV%20nuke.pdf. Accessed November 1, 2011.

George Dyson, son of Freeman Dyson, gives a superb talk on the Orion nuclear propulsion project in this video. He has several diagrams and photographs that have been classified for 55 years. Some of the diagrams he shows have been de-classified, some have not, and some have been reclassified. Available online. URL: http://www.ted.com/talks/george_ dyson_on_project_orion.html. Accessed November 1, 2011.

Hyperion Power Generation provides a Web site that describes the company and its product, a small modular reactor that is manufactured in a factory and shipped to the power plant site on a truck. The quick fact sheet includes a downloadable PDF file containing diagrams and specifications for this interesting development. Available online. URL: http:// www.hyperionpowergeneration.com/. Accessed November 1, 2011.

MNES, or Mitsubishi Nuclear Energy Systems, devotes a section of their Web site to the US-APWR, the advanced pressurized water reactor configured specifically to meet safety standards set by the U.S. Nuclear Regulatory Commission. The site includes a detailed cutaway diagram of the plant, technical specifications, and a history of the US-APWR development. Available online. URL: http://www.mnes-us.com/htm/usapwrdesign. htm. Accessed November 1, 2011.

mPower, a small modular reactor power plant being developed by the Babcock & Wilcox Company, is described on the company Web site. Detailed color cutaway diagrams of the reactor, the containment structure, and the entire power plant are included. Available online. URL: http://www. babcock.com/products/modular_nuclear. Accessed November 1, 2011.

NUREG/CR-6572 is a report prepared by the Brookhaven National Laboratory, entitled "Kalinin VVER-1000 Nuclear Power Station Unit 1 PRA: Procedure Guides for a Probabilistic Risk Assessment." Although this report is slow reading, it is an interesting inside look at a Russian Generation III reactor, as seen through the safety requirements of the U.S. Nuclear Regulatory Commission. Available online. URL: http://pbadupws.nrc. gov/docs/ML0604/ML060450618.pdf. Accessed November 1, 2011.

NuScale Power is developing a modular reactor concept. At their corporate site are descriptions of their new reactor design, along with its advantages and progress with getting a license to build and sell it from the NRC. There are also news stories and a history of NuScale. Available online. URL: http://www.nuscalepower.com. Accessed November 1, 2011.

TerraPower has a Web site that describes the company and the innovative traveling wave reactor. The impressive benefits of the traveling wave technology are explained. For deeper research, the site will guide the user to technical papers that have been written describing this unique reactor concept. Available online. URL: http://www.terrapower.com/Technology/ TravelingWaveReactor.aspx. Accessed November 1, 2011.

Toshiba 4S modular reactors are given a critical review in this 6-page document presented by the Union of Concerned Scientists on December 11, 2006. They point out, for example, that the 4S uses a single control rod, and such a design feature was proven deadly in the SL-1 explosion in Idaho in 1960. This report is a good counterpoint to otherwise glowing reviews of this technology. Available online. URL: http://www.yritwc.org/ Portals/0/PDFs/nuclearreactorletterucs.pdf. Accessed November 1, 2011.

WIPP, the Waste Isolation Pilot Plant, has its own official Web site. Included is access to a collection of documents for WIPP, such as its hazardous waste facility permit, its EPA certification, and a copy of the federal regulations. Available online. URL: http://www.wipp.energy.gov. Accessed November 1, 2011.

Index

Italic page numbers indicate illustrations.